HIGH-DIMENSIONAL COVARIANCE ESTIMATION

WILEY SERIES IN PROBABILITY AND STATISTICS

Established by WALTER A. SHEWHART and SAMUEL S. WILKS

A complete list of the titles in this series appears at the end of this volume.

HIGH-DIMENSIONAL COVARIANCE ESTIMATION

MOHSEN POURAHMADI
Texas A & M University

WILEY

Published by John Wiley & Sons, Inc., Hoboken, New Jersey.
Published simultaneously in Canada.

Library of Congress Cataloging-in-Publication Data:

Pourahmadi, Mohsen.
Modern methods to covariance estimation / Mohsen Pourahmadi, Department of Statistics, Texas A&M University, College Station, TX.
pages cm
Includes bibliographical references and index.
ISBN 978-1-118-03429-3 (hardback)
1. Analysis of covariance. 2. Multivariate analysis. I. Title.
QA279.P68 2013
519.5′38–dc23

2013000326

10 9 8 7 6 5 4 3 2 1

MIX
Paper from responsible sources
FSC® C013604

CONTENTS

PREFACE

The aim of this book is to bring together and present some of the most important recent ideas and methods in high-dimensional covariance estimation. It provides computationally feasible methods and their conceptual underpinnings for sparse estimation of large covariance matrices. The major unifying theme is to reduce sparse covariance estimation to that of estimating suitable regression models using penalized least squares. The framework has the great advantage of reducing the unintuitive and challenging task of covariance estimation to that of modeling a sequence of regressions. The book is intended to serve the needs of researchers and graduate students in statistics and various areas of science, engineering, economics and finance. Coverage is at an intermediate level, familiarity with the basics of regression analysis, multivariate analysis, and matrix algebra is expected.

A covariance matrix, the simplest summary measure of dependence of several variables, plays prominent roles in almost every aspect of multivariate data analysis. In the last two decades due to technological advancements and availability of high-dimensional data in areas like microarray, e-commerce, information retrieval, fMRI, business, and economy, there has been a growing interest and great progress in developing computationally fast methods that can handle data with as many as thousand variables collected from only a few subjects. This situation is certainly not suited for the classical multivariate statistics, but rather calls for a sort of "fast and sparse multivariate methodology."

The two major obstacles in modeling covariance matrices are high-dimensionality (HD) and positive-definiteness (PD). The HD problem is familiar from regression analysis with a large number of covariates where the penalized least squares with the Lasso penalty is commonly used to obtain computationally feasible solutions. However, the PD problem is germane to covariances where one hopes to remove it by infusing regression-based ideas into principal component analysis (PCA), Cholesky decomposition, and Gaussian graphical models (inverse covariance matrices), etc.

The primary focus of current research in high-dimensional data analysis and hence covariance estimation has been on developing feasible algorithms to compute the estimators. There has been less focus on inference and the effort is mostly devoted to establishing consistency of estimators when both the sample size and the number of variables go to infinity in certain manners depending on the nature of sparsity of the model and the data. At present, there appears to be a sort of disconnection between the theory and practice where further research is hoped to bridge the gap. Our coverage follows mostly the recent pattern of research in the HD data literature by focusing more on the algorithmic aspects of the high-dimensional covariance estimation. This is a rapidly growing area of statistics and machine learning, less than a decade old, but has seen tremendous growth in such a short time. Deciding what to include in the first book of its kind is not easy as one does not have the luxury of choosing results that have passed the test of time. My selection of topics has been guided by the promise of lasting merit of some of the existing and freshly minted results, and personal preferences.

The book is divided into two parts. Part I, consisting the first three chapters, deals with the more basic concepts and results on linear regression models, high-dimensional data, regularization, and various models/estimation methods for covariance matrices. Chapter 1 provides an overview of various regression-based methods for covariance estimation, Chapter 2 introduces several examples of high-dimensional data and illustrates the poor performance of the sample covariance matrix and the need for its regularization. A fairly comprehensive review of mathematical and statistical properties of the covariance matrices along with classical covariance estimation results is provided in Chapter 3. Part II is concerned with the modern high-dimensional covariance estimation. It covers shrinkage estimation of covariance matrices, sparse PCA, Gaussian graphical models, and penalized likelihood estimation of inverse covariance matrices. Chapter 6 deals with banding, tapering, and thresholding of the sample covariance matrix or its componentwise penalization. The focus of Chapter 7 is on applications of covariance estimation and singular value decomposition (SVD), to multivariate regression models for high-dimensional data.

The genesis of the book can be traced to teaching a topic course on covariance estimation in the Department of Statistics at the University of Chicago, during a sabbatical in 2001–2002 academic year. I have had the benefits of discussing various topics and issues with many colleagues and students including Anindya Bahdra, Lianfu Chen, Michael Daniels, Nader Ebrahimi, Tanya Garcia, Shuva Gupta, Jianhua Huang, Priya Kohli, Soumen Lahiri, Mehdi Madoliat, Ranye Sun, Adam Rothman, Wei Biao Wu, Dale Zimmerman, and Joel Zinn. Financial support from the NSF in the last decade has contributed greatly to the book project. The editorial staff at John Wiley & Sons and Steve Quigley were generous with their assistance and timely reminders.

MOHSEN POURAHMADI

College Station, Texas
April, 2013

PART I

MOTIVATION AND THE BASICS

CHAPTER 1

INTRODUCTION

Is it possible to estimate a covariance matrix using the regression methodology? If so, then one may bring the vast machinery of regression analysis (regularized estimation, parametric and nonparametric methods, Bayesian analysis, ...) developed in the last two centuries to the service of covariance modeling.

In this chapter, through several examples, we show that sparse estimation of high-dimensional covariance matrices can be reduced to solving a series of regularized regression problems. The examples include sparse principal component analysis (PCA), Gaussian graphical models, and the modified Cholesky decomposition of covariance matrices. The roles of sparsity, the least absolute shrinkage and smooth operator (Lasso) and particularly the soft-thresholding operator in estimating the parameters of linear regression models with a large number of predictors and large covariance matrices are reviewed briefly.

Nowadays, high-dimensional data are collected routinely in genomics, biomedical imaging, functional magnetic resonance imaging (fMRI), tomography, and finance. Let $\boldsymbol{X}$ be an $n \times p$ data matrix where n is the sample size and p is the number of variables. By the *high-dimensional data* usually it is meant that p is bigger than n. Analysis of high-dimensional data often poses challenges which calls for new statistical methodologies and theories (Donoho, 2000). For example, least-squares fitting of linear models and classical multivariate statistical methods cannot handle high-dimensional $\boldsymbol{X}$ since both rely on the inverse of $\boldsymbol{X}'\boldsymbol{X}$ which could be singular or not well-conditioned. It should be noted that increasing n and p each has very different and opposite effects on the statistical results. In general, the focus of multivariate analysis is to make statistical inference about the dependence among variables so

High-Dimensional Covariance Estimation, First Edition. Mohsen Pourahmadi.

that increasing n has the effect of improving the precision and certainty of inference, whereas increasing p has the opposite effect of reducing the precision and certainty. Therefore the level of detail that can be inferred about correlations among variables improves with increasing n but it deteriorates with increasing p.

The dimension reduction and variable selection are of fundamental importance for high-dimensional data analysis. The *sparsity principle* which assumes that only a small number of predictors contribute to the response is frequently adopted as the guiding light in the analysis. Armed with the sparsity principle, a large number of estimation approaches are available to estimate sparse models and select the significant variables simultaneously. The Lasso method introduced by Tibshirani (1996) is one of the most prominent and popular estimation methods for the high-dimensional linear regression models.

Quantifying the interplay between high-dimensionality and sparsity is important in the modern data analysis environment. In the classic setup, usually p is fixed and n grows so that

$$\frac{p}{n} \ll 1.$$

However, in the modern high-dimensional setup where both p and n can grow, estimating accurately a vector β with p parameters is a real challenge. By invoking the *sparsity principle* one assumes that only a small number, say $s(\beta)$, of the entries of β is nonzero and then proceeds in developing algorithms to estimate the nonzero parameters. Of course, it is desirable to establish the statistical consistency of the estimates under a new asymptotic regime where both $n, p \to \infty$. Interestingly, it has emerged from such asymptotic theory that for the consistency to hold in some generic problems, the dimensions n, p of the data and the *sparsity index* of the model must satisfy

$$\frac{\log p}{n} \cdot s(\beta) \ll 1. \tag{1.1}$$

The ratio $\log p/n$ does play a central role in establishing consistency results for variety of covariance estimators proposed for high-dimensional data in recent years (Bühlmann and van de Geer, 2011).

1.1 LEAST SQUARES AND REGULARIZED REGRESSION

The idea of least squares estimation of the regression parameters in the familiar linear model

$$\boldsymbol{Y} = \boldsymbol{X}\beta + \boldsymbol{\varepsilon}, \tag{1.2}$$

has served statistics quite well when the sample size n is large and p is *fixed* and small, say less than 50. The principle of model simplicity or *parsimony* coupled with the techniques of subset, forward, and backward selections have been developed and used to fit such models either for the purpose of describing the data or for its prediction.

The whole machinery of least-squares fails or does not work well for the high-dimensional data where the ubiquitous $\boldsymbol{X}'\boldsymbol{X}$ matrix is not invertible. The traditional remedy is the ridge regression (Hoerl and Kennard, 1970), which replaces the residual sum of squares of errors by its penalized version:

$$Q(\beta) = \|\boldsymbol{Y} - \boldsymbol{X}\beta\|^2 + \lambda \sum_j |\beta_j|^2, \tag{1.3}$$

where $\lambda > 0$ is a penalty controlling the length of the vector of regression parameters. The unique ridge solution is

$$\widehat{\beta}_{\text{ridge}} = (\boldsymbol{X}'\boldsymbol{X} + \lambda \mathbf{I})^{-1}\boldsymbol{X}'\boldsymbol{Y}, \tag{1.4}$$

which amounts to adding λ to the diagonal entries of $\boldsymbol{X}'\boldsymbol{X}$, and then inverting it. The ridge solution works rather well in the presence of multicollinearity and when p is not too large; however, in general it does not induce sparsity in the model. Nevertheless, it points to the fruitful direction of penalizing a norm of the high-dimensional vector of coefficients (parameters) in the model.

In the modern context of high-dimensional data, the standard goals of regression analysis have also shifted toward:

(I) construction of *good predictors* where the actual values of coefficients in the model are *irrelevant*;

(II) giving causal interpretations of the factors and determining which variables are more *important*.

It turns out that regularization is important for both of these goals, but the appropriate magnitude of the regularization parameter depends on which goal is more important for a given problem. Historically, Goal (II) has been the engine of the statistical developments and the thought of irrelevancy of the parameter values was not imaginable. However, nowadays Goal (I) is the primary focus of developing algorithms in the machine learning theory (Bühlmann and van de Geer, 2011). In general, the pair $(\boldsymbol{Y}, \boldsymbol{X})$ is usually modeled nonparametrically as

$$\boldsymbol{Y} = m(\boldsymbol{X}) + \boldsymbol{\varepsilon}, \tag{1.5}$$

with $E(\boldsymbol{\varepsilon}) = 0$ and $m(\cdot)$ a smooth unknown function. Then using a family of basis functions $b_j(\boldsymbol{X})$, $j = 1, 2, \cdots$, $m(\cdot)$ is approximated closely with the sums: $\sum_{j=1}^{p} \beta_j b_j(\boldsymbol{X})$, for a large p, where $\beta = (\beta_1, \cdots, \beta_p)'$ is a vector of coefficients. Of course, when $b_i(\boldsymbol{X}) = \boldsymbol{X}_j$ is the jth column of the design matrix, then the approximation scheme reduces to the familiar linear regression model.

In the high-dimensional contexts, the ridge or ℓ_2 penalty on $||\beta||_2^2 = \sum_{j=1}^{p} \beta_j^2$ has lost some of its attractions to competitors like the Lasso which penalizes large values of the sum $\sum_{j=1}^{p} |\beta_j| = ||\beta||_1$ and hence forces many of the smaller β_j's to be estimated by zero. This zeroing of the coefficients is seen as *model selection* in the sense that only variables with $\beta_j \neq 0$ are included. To this end, perhaps a more natural penalty function is

$$P(\beta) = \sum_{j=1}^{p} I(\beta_j \neq 0) = ||\beta||_0, \tag{1.6}$$

which counts the number of nonzero coefficients. However, the ℓ_0 norm is not an easy function to work with so far as optimization is concerned as it is neither smooth nor convex. Fortunately, the Lasso penalty is the closest convex member of the family of penalty functions of the form $||\beta||_\alpha^\alpha = \sum_{j=1}^{p} |\beta|^\alpha, \alpha > 0$ to (1.6).

1.2 LASSO: SURVIVAL OF THE BIGGER

In this section, we indicate that the Lasso regression which corresponds to replacing the ridge penalty in (1.3) by the ℓ_1 penalty on the coefficients leads to more sparse solutions than the ridge penalty. It forces to zero the smaller coefficients, but keeps the bigger ones around.

The Lasso regression is one of the most popular approaches for selecting significant variables and estimating regression coefficients simultaneously. It corresponds to a penalized least-squares regression with the ℓ_1 penalty on the coefficients (Tibshirani, 1996). Compared to the ridge penalty, it minimizes the sum of squares of residuals subject to a constraint on the sum of absolute values of the regression coefficients:

$$Q(\beta) = \frac{1}{2}||\boldsymbol{Y} - \boldsymbol{X}\beta||^2 + \lambda \sum_{j} |\beta_j|, \tag{1.7}$$

where $\lambda > 0$ is a penalty or tuning parameter controlling the sparsity of the model or the magnitude of the estimates. It is evident that for larger values of the tuning parameter λ, the Lasso estimate $\widehat{\beta}(\lambda)$ obtained by minimizing (1.7) shrinks or forces the regression coefficients toward zero. Since the regularization parameter controls the model complexity, its proper selection is of critical importance in the applications of the Lasso and other penalized least-squares/likelihood methods. In most of what follows in the sequel, it is assumed that λ is fixed and known.

Unlike the closed-form ridge solution in (1.4), due to the nature of the constraint in (1.7), its solution is nonlinear in the responses Y_i's, see (1.11). Fundamental to understanding and computing the Lasso solution is the *soft-thresholding operator*.

Its relevance is motivated using the simple problem of minimizing the function (1.7) for $n = p = 1$, or for a generic observation y and $\boldsymbol{X} = 1$:

$$Q(\beta) = \frac{1}{2}(y - \beta)^2 + \lambda|\beta|, \tag{1.8}$$

for a fixed λ. Note that $|\beta|$ is a differentiable function at $\beta \neq 0$ and the derivative $Q(\cdot)$ with respect to such a β is

$$Q'(\beta) = -y + \beta + \lambda \cdot \text{sign}(\beta) = 0, \tag{1.9}$$

where sign(β) is defined through $|\beta| = \beta \cdot \text{sign}(\beta)$. By convention, its value at zero is set to be zero. The explicit solution of β in terms of y, λ from (1.9) is

$$\widehat{\beta}(\lambda) = \text{sign}(y)(|y| - \lambda)_+, \tag{1.10}$$

where $(x)_+ = x$, if $\quad x > 0 \quad$ and 0, otherwise (see Section 2.5 for details). This simple closed-form solution reveals the following two fundamental characteristics of the Lasso solution:

1. Increasing the penalty λ will shrink the Lasso estimate $\widehat{\beta}(\lambda)$ toward 0. In fact, as soon as λ exceeds $|y|$, the Lasso estimate $\widehat{\beta}(\lambda)$ becomes zero and will remain so thereafter
2. The Lasso solution is *piecewise linear* in the penalty λ (see Fig. 1.1).

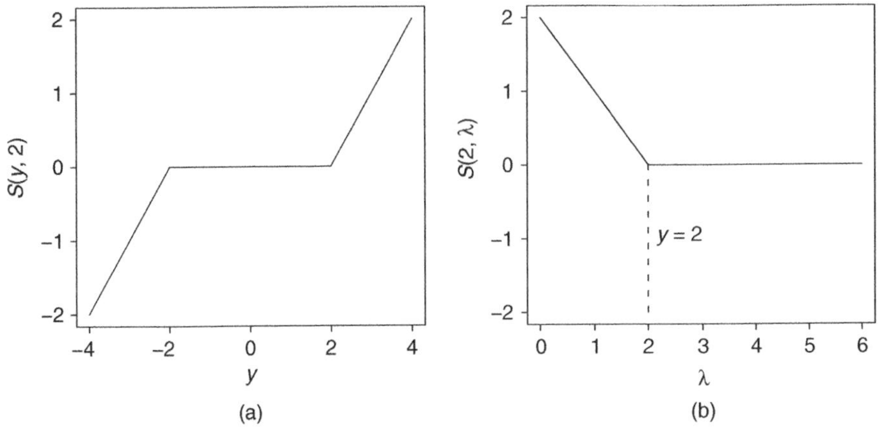

FIGURE 1.1 (a) Plot of the soft-thresholding operator for $\lambda = 2$. (b) Piecewise linearity of Lasso estimator for $y = 2$.

The function (1.10), the building block for solving many penalized regression problems, is called the *soft-thresholding operator*. The following generic notation will be used in the sequel:

$$S(y, \lambda) = \text{sign}(y)(|y| - \lambda)_+, \tag{1.11}$$

where y and λ are real and nonnegative numbers, respectively. There is the related *hard-thresholding function* at threshold λ defined by

$$H(y, \lambda) = yI(|y| > \lambda),$$

where $I(A)$ is the indicator function of the set A, which is also important in the literature of sparse estimation (see Problem 1). Note that these two functions are both nonlinear in y, and have in common the *threshold region* $|y| \leq \lambda$ where the signal is estimated to be zero.

The following example shows the role of the soft-thresholding operator in finding the slope of the simple linear regression with no intercept.

Example 1 (Simple Linear Regression with No Intercept) *Suppose* $\boldsymbol{Y} = (y_1, \cdots, y_n)'$, $\boldsymbol{X} = (x_1, \cdots, x_n)'$, *and consider minimizing:*

$$Q(\beta) = \frac{1}{2}||\boldsymbol{Y} - \beta\boldsymbol{X}||^2 + \lambda|\beta|.$$

For $\lambda = 0$, *the task is to find the least-square estimate of* β *which is given by* $\widehat{\beta} = \sum_{i=1}^n x_i y_i / \sum_{i=1}^n x_i^2$. *In what follows, for simplicity, we assume that the* x*'s are normalized so that* $\sum_{i=1}^n x_i^2 = 1$. *Then, for* $\lambda > 0$, *the Lasso estimate satisfies*

$$Q'(\beta) = -\sum_1^n x_i(y_i - \beta x_i) + \lambda \cdot sign(\beta) = -\sum_{i=1}^n x_i y_i + \beta + \lambda \cdot sign(\beta).$$

Comparing with (1.9), it follows that the Lasso estimate of the slope parameter β *is*

$$\widehat{\beta}(\lambda) = S\left(\sum_{i=1}^n x_i y_i, \lambda\right), \tag{1.12}$$

where the first argument $\widehat{\beta}$ *is the least-squares estimate in a simple linear regression with no intercept.*

The setup of Example 1 and the closed form of its solution play important roles in solving the sparse PCA problem using the singular value decomposition (SVD) of the data matrix $\boldsymbol{X}$ described later in Section 1.4 and Chapter 4.

For a general design matrix $\boldsymbol{X}$, the computational details of the Lasso solution are more involved and will be discussed in Chapter 2. A fast method to compute it is the *coordinate descent algorithm* which minimizes (1.7) over one β_j at a time with the others kept fixed. It then cycles through all the parameters until its convergence (see Example 11, Friedman et al., 2008; Bühlmann and van de Geer, 2011). For an excellent retrospective review of the basic idea of Lasso, its history, computational developments, and generalizations since the publication of the original paper in 1996 see Tibshirani (2011).

Due to its ability to force to zero smaller parameters, the Lasso penalty and method have been applied to sparse PCA, sparse estimation of covariance and precision matrices, Gaussian graphical models, and multivariate regression. Some of these applications are reviewed briefly in this chapter. Many other sparsity-inducing penalties such as the smoothly clipped absolute deviation (SCAD) (Fan and Li, 2001), the elastic net (Zou and Hastie, 2005), and the adaptive Lasso (Zou, 2006), reviewed in Chapter 2, can be used in the context of covariance estimation. However, for clarity and simplification in notation throughout this book we rely on the Lasso penalty.

1.3 THRESHOLDING THE SAMPLE COVARIANCE MATRIX

For a column-centered $n \times p$ data matrix $\boldsymbol{X}$, it is known that the sample covariance matrix $\mathbf{S} = \frac{1}{n}\boldsymbol{X}'\boldsymbol{X}$ performs poorly in high dimensions. In analogy with the nonparametric estimation of mean vectors (Johnstone, 2011), when the population covariance matrix is sparse one may regularize $\mathbf{S}$ by applying the hard- or soft-thresholding operator (1.11) to it elementwise.

More formally, a soft-thresholded covariance estimator is defined by

$$\widehat{\boldsymbol{\Sigma}}_\lambda = S(\mathbf{S}, \lambda),$$

where $\lambda > 0$ is a tuning parameter and $S(\cdot, \lambda)$ acts componentwise. Interestingly, such a regularized covariance estimator is the solution of the following convex optimization problem:

$$\begin{aligned}
\widehat{\boldsymbol{\Sigma}}_\lambda &= \underset{\boldsymbol{\Sigma}}{\operatorname{argmin}} \left\{ \frac{1}{2} ||\boldsymbol{\Sigma} - \mathbf{S}||_F^2 + \lambda |\boldsymbol{\Sigma}|_1 \right\}, \\
&= \underset{\boldsymbol{\Sigma}}{\operatorname{argmin}} \left\{ \frac{1}{2} \sum_{i=1}^{p} \sum_{j=1}^{p} (\sigma_{ij} - s_{ij})^2 + \lambda \sum_{i=1}^{p} \sum_{j \neq i}^{p} |\sigma_{ij}| \right\}, \qquad (1.13) \\
&= \underset{\boldsymbol{\Sigma}}{\operatorname{argmin}} \sum_{i=1}^{p} \sum_{j \neq i}^{p} \left\{ \frac{1}{2} (\sigma_{ij} - s_{ij})^2 + \lambda |\sigma_{ij}| \right\},
\end{aligned}$$

where $|| \cdot ||_F$ stands for the Frobenius norm and $| \cdot |_1$ stands for the ℓ_1 norm of the nondiagonal elements of its matrix argument. Note that the optimization problem above is the same as minimizing for each $i \neq j$, the expression inside the last curly bracket which is of the form of (1.8) and hence amounts to soft-thresholding the entry s_{ij}.

It is known that under some regularity conditions such a simple, sparse covariance estimator is consistent in high dimensions (see Chapter 6; Bickel and Levina, 2008b; Rothman et al., 2009). A consequence of the consistency of the thresholded estimator is that its positive-definiteness is guaranteed in the asymptotic setting with high probability. However, the actual estimator is not necessarily a positive-definite matrix in real data analysis. This is illustrated clearly by Xue et al. (2012) using the Michigan lung cancer data with $n = 86$ tumor samples from patients with lung adenocarcinomas and 5217 gene expression values for each sample. Randomly choosing p genes ($p = 200, 500$) they obtained the $p \times p$ soft-thresholding sample correlation matrix for these genes, and repeated the process 10 times for each p, each time the thresholding parameter λ was selected via a five-fold cross-validation. It was found that none of the soft-thresholding estimators were positive-definite. On the average, there were 22 and 124 negative eigenvalues for the soft-thresholding estimator for $p = 200$ and 500, respectively.

To deal with the lack of positive-definiteness, one popular solution is to utilize the eigendecomposition of $\mathbf{S}$ and project it into the convex cone of positive-definite matrices. Let

$$\mathbf{S} = \sum_{i=1}^{p} \widehat{\lambda}_i \widehat{e}_i \widehat{e}'_i,$$

be its eigendecomposition then a positive semidefinite estimator can be obtained by setting

$$\tilde{\mathbf{S}}^+ = \sum_{i=1}^{p} \max(\widehat{\lambda}_i, 0) \widehat{e}_i \widehat{e}'_i.$$

Unfortunately, this strategy does not work well for sparse covariance matrix estimation, because the projection destroys the sparsity pattern of $\mathbf{S}$. For the Michigan lung cancer data, after the semidefinite projection, the soft-thresholding estimator has no zero entry.

In order to simultaneously achieve sparsity and positive-definiteness, a natural approach is to add a positive-definiteness constraint to the objective function (see Chapter 6 and Rothman, 2012).

1.4 SPARSE PCA AND REGRESSION

In this section, we present a simple regression-based procedure for regularizing the principal components (PCs) using the close connections among the SVD of the column-centered data matrix $\boldsymbol{X}$, the eigen-decomposition of its sample covariance matrix $\mathbf{S}$, and the low-rank approximation property of SVD in the Frobenius norm of a matrix (see Section 4.3).

One of the most popular techniques in multivariate statistics is the principal component analysis (PCA) (see Section 3.5). It is used for dimension reduction, data visualization, data compression, and information extraction by relying on the first few eigenvectors of the sample covariance matrix. For a data matrix $\boldsymbol{X}$, PCA finds sequentially unit vectors $\mathbf{v}_1, \cdots, \mathbf{v}_p$ that maximize $\mathbf{v}'\boldsymbol{X}'\boldsymbol{X}\mathbf{v}$ subject to $\mathbf{v}_{i+1}$ being orthogonal to the previous vectors $\mathbf{v}_1, \cdots, \mathbf{v}_i$. These are the (sample) *principal component (PC) loadings* and the vectors $\boldsymbol{X}\mathbf{v}_i$ are the *principal components (PCs)*. A major problem in the high-dimensional data situations is that the classical PCs have poor statistical properties in the sense that the sample eigenvectors are not consistent estimators of their population counterparts (Johnstone and Lu, 2009). Thus, it is necessary to regularize them in some manners.

We focus on the recent growing interest in developing sparse PCA via the SVD of a data matrix using the following simple matrix algebra facts:

1. The normalized right singular vectors of the column-centered data matrix $\boldsymbol{X}$ and the normalized eigenvectors of the sample covariance matrix $\mathbf{S} = 1/n\boldsymbol{X}'\boldsymbol{X}$ are the same (see Section 4.3).
2. The best rank-one approximation to a data matrix $\boldsymbol{X}$ in the Frobenius norm is given by a constant multiple of $\boldsymbol{u}_1\mathbf{v}_1'$ where $\boldsymbol{u}_1$ and $\mathbf{v}_1$ are its first left and right singular vectors.

To regularize an approximant $\boldsymbol{u}\mathbf{v}'$, we restrict $\boldsymbol{u}$ to have unit length, that is $||\boldsymbol{u}||_2^2 = \sum_{i=1}^n u_i^2 = 1$, and assume that $\mathbf{v}$ is sparse with many zeros. Then, imposing a Lasso penalty on the vector of PC loadings the task is to minimize the penalized objective function:

$$\|\boldsymbol{X} - \boldsymbol{u}\mathbf{v}\|_F^2 + \lambda \sum_{j=1}^{p} |v_j| = \sum_j \left\{ \sum_i (x_{ij} - u_i v_j)^2 + \lambda |v_j| \right\}, \tag{1.14}$$

subject to $||\boldsymbol{u}||_2 = 1$, where v_j is the jth entry of the vector of PC loadings.

The full computational advantage of the *regression-based approach* to PCA emerges by noting that the inner sum in (1.14), for each j, is precisely the penalized regression problem studied in Example 1. Thus, for a fixed $\boldsymbol{u}$, the solution vector $\mathbf{v}$ is the (penalized) regression coefficient when the p columns of $\boldsymbol{X}$ are regressed on $\boldsymbol{u}$ one at a time (Shen and Huang, 2008). Of course, for a fixed $\mathbf{v}$ a similar and much simpler statement is true about computing $\boldsymbol{u}$. These useful facts are summarized in the following lemma:

Lemma 1 (Regression-based SVD)

(a) For a fixed $\mathbf{v}$, *the minimizer of (1.14) with* $||\boldsymbol{u}||_2 = 1$ *is given by*

$$\boldsymbol{u} = \frac{\boldsymbol{X}\mathbf{v}}{||\boldsymbol{X}\mathbf{v}||_2}.$$

(b) For a fixed $\boldsymbol{u}$*, the minimizer of (1.14) is given by*

$$\mathbf{v} = S\left(\boldsymbol{X}'\boldsymbol{u}, \lambda\right),$$

where the soft-thresholding operator acts componentwise on the vector $\boldsymbol{X}'\boldsymbol{u}$*.*

Lemma 1 suggests the following iterative procedure for solving the optimization problem (1.14):

The Shen and Huang (2008) Algorithm for Minimizing (1.14):

---

1. Initialize $\mathbf{v}$ with $\|\mathbf{v}\|_2 = 1$
2. Iterate until convergence:
   (a) $\boldsymbol{u} \leftarrow \frac{\boldsymbol{X}\mathbf{v}}{\|\boldsymbol{X}\mathbf{v}\|_2}$,
   (b) $\mathbf{v} \leftarrow S(\boldsymbol{X}'\boldsymbol{u}, \lambda)$.

One may initialize $\mathbf{v}$ using the first right singular vector of the standard SVD of $\boldsymbol{X}$. In the literature of PCA, it is common for $\mathbf{v}$ to have unit length for the sake of identifiability. This can be done at the convergence by normalizing the PC loadings to have $||\mathbf{v}||_2 = 1$.

Interestingly, the previous algorithm is closely connected to the *power or alternating least-squares (ALS) method* for computing the standard SVD of a data matrix. In fact, for $\lambda = 0$ since $S\left(\boldsymbol{X}'\boldsymbol{u}, 0\right) = \boldsymbol{X}'\boldsymbol{u}$, it reduces to the *power method* for computing the eigenvectors (see Golub and Van Loan, 1996; Gabriel and Zamir, 1979). Starting with $\mathbf{v}^{(0)}$ it is seen from the algorithm that at the end of the kth iteration

$$\mathbf{v}^{(k)} = \frac{(\boldsymbol{X}'\boldsymbol{X})^k \mathbf{v}^{(0)}}{||(\boldsymbol{X}'\boldsymbol{X})^k \mathbf{v}^{(0)}||_2}, \tag{1.15}$$

which computes the largest eigenvector of $\boldsymbol{X}'\boldsymbol{X}$ or the leading right singular vector of $\boldsymbol{X}$. This connection is made more transparent and discussed further in Section 4.3.

Thus far, our focus has been on estimating a covariance matrix $\boldsymbol{\Sigma}$. In some applications, such as model-based clustering, classification, and regression, the need for the precision matrix $\boldsymbol{\Sigma}^{-1}$ is stronger than that for $\boldsymbol{\Sigma}$ itself. In the standard multivariate statistics, the former is usually computed from the latter. However, since inversion tends to destroy sparsity and may introduce noise for large p, it is advantageous to treat estimating a sparse covariance matrix and sparse precision matrix as completely separate problems. Furthermore, inversion of a $p \times p$ matrix requires $\mathcal{O}(p^3)$ operations which in the high-dimensional situation is computationally expensive and can be avoided by using the techniques explained next.

1.5 GRAPHICAL MODELS: NODEWISE REGRESSION

Gaussian graphical models explore and display conditional dependence relationships among Gaussian random variables. A connection between conditional independence of variables in a Gaussian random vector and graphs is made by identifying the graph's nodes with random variables and translating the graph's edges into a parametrization that relates to the precision matrix of a multivariate normal distribution.

In this section, an overview of a regression-based approach to inducing sparsity in the precision matrix is given. Sparsity of the precision matrix is of great interest in the literature of Gaussian graphical models (cf. Whittaker, 1990; Hastie et al., 2009, Chap. 17). The key idea is that in a multivariate normal random vector $\boldsymbol{Y}$, the conditional dependencies among its entries are related to the off-diagonal entries of its precision matrix $\boldsymbol{\Sigma}^{-1} = (\sigma^{ij})$. More precisely, the variables i and j are conditionally independent given all other variables, if and only if $\sigma^{ij} = 0$, so that the problem of estimating a Gaussian graphical model is equivalent to estimating a precision matrix.

To cast the problem in the language of linear regression, let $\widehat{Y}_i = \sum_{j \neq i} \beta_{ij} Y_j$ be the linear least-squares predictor of Y_i based on the rest of the components of $\boldsymbol{Y}$ and $\varepsilon_i = Y_i - \widehat{Y}_i$ be its prediction error. Then, writing

$$Y_i = \sum_{j \neq i} \beta_{ij} Y_j + \varepsilon_i, \tag{1.16}$$

we show in Section 5.2 that the coefficients $\beta_{i,j}$ of Y_i based on the remaining variables and the corresponding prediction error variances are given by:

$$\beta_{i,j} = -\frac{\sigma^{ij}}{\sigma^{ii}},\ j \neq i, \quad \mathrm{Var}(Y_i | Y_j, j \neq i) = \frac{1}{\sigma^{ii}},\ i = 1, \cdots, p. \tag{1.17}$$

This confirms that σ^{ij}, the (i, j)th entry of the precision matrix is, up to a scalar, the coefficient of variable j in the multiple regression of the node or variable i on the rest. As such, each $\beta_{i,j}$ is an unconstrained real number with $\beta_{j,j} = 0$, but note that $\beta_{i,j}$ is not necessarily symmetric in (i, j).

Now, for each node i one may impose an ℓ_1 penalty on the regression coefficients in (1.16). The idea of nodewise Lasso regression, proposed first in Meinshausen and Bühlmann (2006), has been the source of considerable research in sparse estimation of precision matrices leading to the penalized normal likelihood estimation and the graphical Lasso (Glasso) algorithm discussed in Chapter 5.

1.6 CHOLESKY DECOMPOSITION AND REGRESSION

In this section, the close connection between the modified Cholesky decomposition of a covariance matrix and the idea of regression is reviewed when the variables are ordered or there is a notion of distance between variables, as in time series, longitudinal and functional data, and spectroscopic data.

In what follows, it is assumed that $\boldsymbol{Y}$ is a time-ordered random vector with mean zero and a positive-definite covariance matrix $\boldsymbol{\Sigma}$. Let $\widehat{Y}_t$ be the linear least-squares predictor of Y_t based on its predecessors $Y_{t-1}, \ldots, Y_1$ and $\varepsilon_t = Y_t - \widehat{Y}_t$ be its prediction error with variance $\sigma_t^2 = \text{var}(\varepsilon_t)$. Then, there are unique scalars ϕ_{tj} so that

$$Y_t = \sum_{j=1}^{t-1} \phi_{tj} Y_j + \varepsilon_t, \; t = 1, \cdots, p, \tag{1.18}$$

and the regression coefficients ϕ_{tj} can be computed using the covariance matrix as shown in Section 3.6. Let $\varepsilon = (\varepsilon_1, \cdots, \varepsilon_p)'$ be the vector of successive uncorrelated prediction errors with

$$\text{cov}(\varepsilon) = \text{diag}\left(\sigma_1^2, \cdots, \sigma_p^2\right) = D.$$

Then, (1.18) can be rewritten in matrix form as $T\boldsymbol{Y} = \varepsilon$, where T is the following unit lower triangular matrix with $-\phi_{tj}$ as its (t, j)th entry:

$$T = \begin{pmatrix} 1 & & & & \\ -\phi_{21} & 1 & & & \\ -\phi_{31} & -\phi_{32} & 1 & & \\ \vdots & & & \ddots & \\ -\phi_{p1} & -\phi_{p2} & \cdots & -\phi_{p,p-1} & 1 \end{pmatrix}. \tag{1.19}$$

Now, computing the covariance

$$\text{cov}(\varepsilon) = \text{cov}(T\boldsymbol{Y}) = T\boldsymbol{\Sigma}T',$$

one arrives at the decompositions

$$T\boldsymbol{\Sigma}T' = D, \boldsymbol{\Sigma}^{-1} = T'D^{-1}T. \tag{1.20}$$

Thus, both $\boldsymbol{\Sigma}$ and $\boldsymbol{\Sigma}^{-1}$ are diagonalized by unit lower triangular matrices; we refer to (1.20) as the modified Cholesky decompositions of the covariance and its precision matrix, respectively.

Since the ϕ_{ij}'s in (1.18) are simply the regression coefficients computed from an unstructured covariance matrix, these coefficients along with $\log \sigma_t^2$ are unconstrained (Pourahmadi, 1999). From (1.18) and the subsequent results, it is evident that the task of modeling a covariance matrix is reduced to that of a sequence of p varying-coefficient and varying-order regression models. Thus, one can bring the whole regression analysis machinery to the service of the unintuitive and challenging task of modeling covariance matrices. In particular, the Lasso penalty can be imposed on the regression coefficients in (1.18) as in Huang et al. (2006). An important consequence of (1.20) is that for any estimate $(\widehat{T}, \widehat{D})$ of the Cholesky factors, the estimated covariance and precision matrix, $\widehat{\boldsymbol{\Sigma}}^{-1} = \widehat{T}'\widehat{D}^{-1}\widehat{T}$, are guaranteed to be positive-definite.

1.7 THE BIGGER PICTURE: LATENT FACTOR MODELS

A broader framework for covariance matrix reparameterization and modeling is the class of *latent factor models* (Anderson, 2003) or *linear mixed models* (Searle et al., 1992). It has the goal of finding a few common (latent, random) factors that may explain the covariance matrix of the data.

For a random vector $\boldsymbol{Y} = (Y_1, \ldots, Y_p)'$ with mean $\boldsymbol{\mu}$ and covariance matrix $\boldsymbol{\Sigma}$, the classical factor model postulates that entries of $\boldsymbol{Y}$ are linearly dependent on a few *unobservable random variables* $F = (f_1, \ldots, f_q)'$, $q \ll p$, and p additional noises $\boldsymbol{\varepsilon} = (\varepsilon_1, \ldots, \varepsilon_p)'$. The f_i's are called *the common factors* and the ε_i's are the *idiosyncratic errors*. More precisely, a *factor model* for $\boldsymbol{Y}$ is

$$
\begin{array}{rcccccccc}
Y_1 - \mu_1 & = & \ell_{11} f_1 & + & \cdots & + & \ell_{1q} f_q & + & \varepsilon_1, \\
Y_2 - \mu_2 & = & \ell_{21} f_1 & + & \cdots & + & \ell_{2q} f_q & + & \varepsilon_2, \\
 & \vdots & & & & & & & \\
Y_p - \mu_p & = & \ell_{p1} f_1 & + & \cdots & + & \ell_{pq} f_q & + & \varepsilon_p,
\end{array}
$$

or in matrix notation

$$
\boldsymbol{Y} - \boldsymbol{\mu} = LF + \varepsilon, \tag{1.21}
$$

where $L = (\ell_{ij})$ is the $p \times q$ matrix of *factor loadings* (see Section 3.7 for more details).

Algebraically, the factor model (1.21) amounts to decomposing a $p \times p$ covariance matrix as

$$
\boldsymbol{\Sigma} = L\Lambda L' + \Psi, \tag{1.22}
$$

where L is a $p \times q$ matrix, Λ and Ψ are usually $q \times q$ and $p \times p$ diagonal matrices, respectively. Note that the factor model swaps $\boldsymbol{\Sigma}$ by the triplet (L, Λ, Ψ) which leads to a considerable reduction in the number of covariance parameters if q is small relative to p. The generality of (1.22) becomes more evident by choosing the components of the quadruple (q, L, Λ, Ψ) in various ways as indicated below:

1. *Spectral decomposition or PCA* amounts to choosing $q = p$, $\Psi = 0$, L as the orthogonal matrix of eigenvectors, and Λ as the diagonal matrix of eigenvalues of the covariance matrix.
2. *Spiked covariance models* (Paul, 2007; Johnstone and Lu, 2009) are special orthogonal factor models obtained by choosing $q < p$, Λ the identity matrix,

and $\Psi = \sigma^2 I$. In this case, writing $L = (L_1, \cdots, L_q)$ where the columns L_j's are orthogonal with decreasing norms

$$||L_1||_2 > ||L_2||_2 > \cdots > ||L_q||_2,$$

leads to the decomposition

$$\Sigma = \sum_{j=1}^{q} L_j L_j' + \sigma^2 I, \tag{1.23}$$

where the L_j's are the ordered eigenvectors of the population covariance matrix. The spiked covariance models are ideal for studying consistency of high-dimensional standard PCA. For example, it will be shown in Chapter 4 that the sparse PCA is consistent for the spiked models if the leading L_j's are sparse or concentrated vectors.

3. *Modified Cholesky decomposition* amounts to taking L to be a unit lower triangular matrix, Λ the diagonal matrix of innovation variances, and $\Psi = 0$.

A possible drawback of the standard latent factor models is that the decomposition (1.22) is not unique, see (3.40) and the subsequent discussions. A common approach to ensure uniqueness is to constrain the factor loadings matrix L to be block lower triangular matrix with strictly positive diagonal elements (Geweke and Zhou, 1996). Unfortunately, such a constraint induces order dependence among the responses or the variables in $\boldsymbol{Y}$ as in the Cholesky decomposition. However, for certain tasks such as prediction, identifiability of a unique decomposition may not be necessary.

The *precision matrix* is needed in variety of statistical applications (Anderson, 2003) and in portfolio management (Fan and Lv, 2008). For the latent factor model (1.22), it can be shown that

$$\Sigma^{-1} = \Psi^{-1} - \Psi^{-1} L(\Lambda + L'\Psi^{-1}L)^{-1} L'\Psi^{-1}, \tag{1.24}$$

which involves inversion of smaller $q \times q$ matrices as Ψ is assumed to be diagonal in the standard factor models. This assumption may not be valid in some economics and financial studies (Chamberlain and Rothschild, 1983) in which case it is more appropriate to assume that Ψ is sparse.

When working with a collection of covariance matrices $\Sigma_1, \cdots, \Sigma_K$ sharing some similarities, the factor model framework can be used to model them as

$$\Sigma_k = \Sigma + L\Lambda_k L', k = 1, \cdots, K, \tag{1.25}$$

where Σ is a common covariance matrix, L is a $p \times q$ full rank matrix, and Λ_k is a $q \times q$ nonnegative definite matrix which is not necessarily diagonal. The maximum likelihood estimation of such a family of reduced covariance matrices is developed in Schott (2012). An important example of a collection of covariance matrices arises in multivariate time-varying volatility $\mathbf{\Sigma}_t, t = 1, \cdots, T$ for p assets over T time periods. It has many applications in finance including asset allocation and risk management. Here, using a standard factor model one writes

$$\mathbf{\Sigma}_t = L_t \Lambda_t L_t' + \Psi_t, t = 1, \cdots, T, \tag{1.26}$$

where the $p \times q$ matrix of factor loadings L_t is block lower triangular matrix with diagonal elements equal to one, Λ_t, Ψ_t are the diagonal covariance matrices of the common and specific factors, respectively. In the literature of finance, (1.26) is referred to as the *factor stochastic volatility* model.

For the versatility of the latent factor models in dealing with large covariances arising from spatial data, financial data, etc., see Fox and Dunson (2011). They propose a Bayesian nonparametric approach to multivariate covariance regression where L in (1.22) and hence the covariance matrix are allowed to change flexibly with covariates or predictors. The approach readily scales to high-dimensional datasets. The JRSS B discussion paper by Fan et al. (2013) offers the most recent and diverse perspectives of research on the roles of factor models in covariance estimation.

1.8 FURTHER READING

An excellent unified view of regularization methods in statistics is given by Bickel and Li (2006). All regularization methods require a tuning parameter. Choosing it too small keeps many variables in the model and retains too much variance in estimation and prediction, but choosing it too large knocks out many variables and introduces too much bias. Thus, controlling the bias–variance tradeoff is the key in selecting the tuning parameter. Cross-validation is the most popular method which requires data splitting and have been only justified for low-dimensional settings. In the following, an outline of some recent progress and relevant references are provided for interested readers.

Theoretical justifications of most methods for tuning parameter selection are usually based on oracle choices of some related parameters (like the noise variance) that cannot be implemented in practice. Choosing the regularization parameter in a data-dependent way remains a challenging problem in a high-dimensional setup. Since the introduction of the degrees of freedom for Lasso regression (Zou et al., 2007), it seems some older techniques such as the C_p-statistic, Akaike information criterion (AIC), Bayesian information criterion (BIC) are becoming popular in the high-dimensional context. Significant progress has been made recently on developing likelihood-free regularization selection techniques like subsampling (Meinshausen and Bühlmann, 2010), which are computationally expensive and still may lack theoretical guarantees.

There is a growing desire and tendency to avoid data-based methods of selecting the tuning parameter by exploiting certain optimal value of the asymptotic thresholding parameter from the Gaussian sequence models (Johnstone, 2011; Yang et al., 2011;

Cai and Liu, 2011). A new procedure (Liu and Wang, 2012) for estimating high-dimensional Gaussian graphical models, called TIGER (tuning-insensitive graph estimation and regression), enjoys a *tuning-insensitive property*, in the sense that it automatically adapts to the unknown sparsity pattern and is asymptotically tuning-free. In finite sample settings, one only needs to pay minimal attention to tuning the regularization parameter. Its main idea, like that in Glasso, is to estimate the precision matrix one column at a time. For each column, the computation is reduced to a sparse regression problem where unlike the existing methods, the TIGER solves the sparse regression problem using the recent SQRT-Lasso method proposed by Belloni et al. (2012). The main advantage of the TIGER over the existing methods is its asymptotic tuning-free property, which allows one to use the entire data to efficiently estimate the model.

PROBLEMS

1. For $y \in R$, $\lambda > 0$, consider the objective function

$$f_\lambda(\beta) = \beta^2 - 2y\beta + p_\lambda(|\beta|),$$

where $p_\lambda(\cdot)$ is a penalty function. Let $\hat{\beta}$ be a minimizer of the objective function.

(a) For the ridge penalty $p_\lambda(|\beta|) = \lambda\beta^2$, show that the minimizer of the objective function is

$$R(y, \lambda) = \frac{y}{1+\lambda}.$$

(b) For $p_\lambda(|\beta|) = \lambda^2 I(|\beta| \neq 0)$, plot $f_2(\beta)$ when $y = 3$. Is the function continuous? Differentiable? Convex?

(c) Show that the minimizer of $f_\lambda(y)$ is the hard-thresholding function

$$H(y, \lambda) = yI(|y| > \lambda).$$

(d) Repeat (c) for the penalty function $p_\lambda(|\beta|) = 2\lambda|\beta|$ and show that the minimizer of the objective function is the soft-thresholding function $S(y, \lambda)$.

(e) Plot $H(y, \lambda)$ and $S(y, \lambda)$ as functions of the data y for $\lambda = 1, 5, 7$. What are the similarities and differences between these two penalized least-squares estimators?

2. *Soft–hard-thresholding function*: Plot the function

$$SH(y, \lambda_1, \lambda_2) = \begin{cases} 0 & \text{if } |y| \leq \lambda_1, \\ \text{sign}(y)\frac{\lambda_2(|y|-\lambda_1)}{\lambda_2-\lambda_1} & \text{if } \lambda_1 \leq |y| \leq \lambda_2, \\ y & \text{if } |y| > \lambda_2, \end{cases}$$

for $\lambda_1 = 1, \lambda_2 = 2$. Compare its similarities and differences with the hard- and soft-thresholding functions.

3. *Smoothly Clipped Absolute Deviation (SCAD) Penalty* (Fan and Li, 2001) is given by

$$p_\lambda(|\beta|) = 2\lambda|\beta|I(|\beta| \leq \lambda) \\ -\frac{\beta^2 - 2a\lambda|\beta| + \lambda^2}{a-1} I(\lambda < |\beta| \leq a\lambda) + (a+1)\lambda^2 I(|\beta| > a\lambda),$$

where $a > 2$ is a second tuning parameter (here it is fixed at 3.7).

(a) Plot the function $p_2(|\beta|)$. Is the function convex (concave)?

(b) Show that the minimizer of the objective function in Problem 1 with the SCAD penalty function is

$$\hat{\beta}(y, \lambda) = \begin{cases} \text{sign}(y)(|y| - \lambda)_+ & \text{if } \ |y| \leq 2\lambda, \\ \{(a-1)y - \text{sign}(y)a\lambda\}/(a-2) & \text{if } \ 2\lambda < |y| \leq a\lambda, \\ y & \text{if } \ |y| > a\lambda. \end{cases}$$

(c) What are the similarities and differences between the penalized least-squares estimators $S(y, \lambda)$ and $\hat{\beta}(y, \lambda)$ corresponding to the soft-thresholding and SCAD penalties?

(d) Compare the SCAD penalty function with the following simpler penalty functions:

$$p_\lambda(|\beta|) = \lambda|\beta|/(1 + |\beta|), \quad p'_\lambda(|\beta|) = (a\lambda - |\beta|)_+/a,$$

where $p'(\cdot)$ stands for the derivative of the function $p(\cdot)$.

4. Consider the problem of minimizing

$$||X - du\mathbf{v}'||_F^2 + d(\lambda_u||u||_1 + \lambda_v||\mathbf{v}||_1), \text{ subject to } \quad ||u||_2 = ||\mathbf{v}||_2 = 1, \quad (1.27)$$

with respect to $(d, u, \mathbf{v})$ where λ_u, λ_v are fixed tuning parameters.

(a) Show that for u fixed, minimizing the function (1.27) with respect to $(d, \mathbf{v})$ is equivalent to minimizing with respect to $\tilde{\mathbf{v}} = d\mathbf{v}$ the function

$$||X - u\tilde{\mathbf{v}}'||_F^2 + \lambda_v \sum_{j=1}^{p} |\tilde{v}_j| = ||Y - (I_p \otimes u)\tilde{\mathbf{v}}||^2 + \lambda_v \sum_{j=1}^{p} |\tilde{v}_j|, \quad (1.28)$$

where $Y = (\mathbf{x}_1', \cdots, \mathbf{x}_p')' \in R^{np}$ with $\mathbf{x}_j$ being the jth column of X and $\otimes$ is the Kronecker product.

(b) Observe that the right-hand side of (1.28) is a Lasso penalized regression problem with the response $\boldsymbol{Y}$, the design matrix $I_p \otimes \boldsymbol{u}$, and the vector of regression coefficients $\tilde{\mathbf{v}}$. Given the first two and λ_v, what algorithm would you use to compute the Lasso estimate of $\tilde{\mathbf{v}}$? Why?

(c) Find the analog of (1.28) for fixed $\mathbf{v}$ when minimizing the function (1.27) with respect to $(d, \boldsymbol{u})$.

CHAPTER 2

DATA, SPARSITY, AND REGULARIZATION

This chapter presents a few examples of high-dimensional data matrices. We highlight the poor performance of the sample covariance matrix in high-dimensional situations and discuss some early remedies such as shrinking its eigenvalues toward a central value. The notions of *sparsity* and *regularization* are discussed in more detail compared to the previous chapter. These are the minimal ingredients needed for introducing effective procedures for estimation and prediction in the high-dimensional data situations. The close interplay between sparsity and regularization, and embedding covariance estimation within the context of mean modeling are reviewed. The rest of the chapter is devoted to presenting additional properties of the Lasso estimates and its various improvements.

Model sparsity usually refers to the presence of many zeros in the high-dimensional parameter objects. Prominent examples include sparse regression models, sparse covariance, and precision matrices. Data sparsity occurs when variables are highly redundant and their underlying structures can be described by a few derived variables or linear combinations of the original variables as, for example, in PCA. Prominent examples of data sparsity are approximate collinearity, highly correlated data, and perhaps data with many missing values as in the Netflix data, see Example 4. Heuristically, the two notions of model and data sparsity are related to each other. For example, an intrinsically low-dimensional data may barely have enough information for accurate estimation of parameters of even a sparse model, let alone nonsparse models. In the high-dimensional data situations, one would take advantage of sparsity using regularization methods which penalize nonsparse models. Important examples of regularization methods considered in this book are the Lasso regression with many

High-Dimensional Covariance Estimation, First Edition. Mohsen Pourahmadi.

variables; nodewise Lasso regression in the high-dimensional Gaussian graphical models and regularized covariance estimation methods such as banding, tapering, and thresholding, and penalized likelihood estimation.

2.1 DATA MATRIX: EXAMPLES

Data can have diverse forms and shapes such as vectors, curves, and images. By digitizing the latter two, we focus here on a common data matrix structure which allows to describe a great number of applications with great economy. A data matrix is a rectangular array with n rows and p columns where the rows correspond to different observations or individuals and the columns correspond to different variables.

In many applications, it is common to collect data on p variables, $Y_1, Y_2, \ldots, Y_p$ for n different subjects or cases. The measurements for the ith subject is usually denoted by a vector $\boldsymbol{Y}_i = (Y_{i1}, Y_{i2}, \ldots, Y_{ip})'$, which can be viewed as the ith row of the following $n \times p$ table or data matrix:

		Variables					
		1	2	$\cdots$	j	$\cdots$	p
	1	Y_{11}	Y_{12}	$\cdots$	Y_{1j}	$\cdots$	Y_{1p}
	2	Y_{21}	Y_{22}	$\cdots$	Y_{2j}	$\cdots$	Y_{2p}
Units	$\vdots$	$\vdots$	$\vdots$		$\vdots$		$\vdots$
	i	$(Y_{i1}$	Y_{i2}	$\cdots$	Y_{ij}	$\cdots$	$Y_{ip})' = \boldsymbol{Y}_i$
	$\vdots$	$\vdots$	$\vdots$		$\vdots$		$\vdots$
	n	Y_{n1}	Y_{n2}	$\cdots$	Y_{nj}	$\cdots$	Y_{np}

We use the notation $\boldsymbol{Y}$ to denote this data matrix. When working with several data matrices, boldface capital letters like $\boldsymbol{X}, \boldsymbol{Z}$ will be used to denote such data matrices. Depending on the subject matter, correlation structure among rows and columns, relative sizes of n and p, and order among the variables, a data matrix may be referred to simply as a multivariate data, time series data, longitudinal data, functional data, spatial data, transposable data, etc.

1. **Multivariate Data**: Usually a number of characteristics of interest, like blood pressure, height, weight, income, etc., are measured for each of n subjects in a sample; the rows of $\boldsymbol{Y}$ are independent of each other (see Example 2).
2. **Functional Data**: The underlying objects of interest are curves. Strictly speaking p is infinity, but when the curve is digitized, then p is finite and large relative to n. The rows are usually independent (Ramsey and Silverman, 2005).
3. **Spatial or Transposable Data**: n, p are large, both rows and columns are correlated (Cressie, 2003; Allen & Tibshirani, 2010).
4. **Time Series Data**: $n = d$ is fixed and p is large where $d \geq 1$ variables are measured over time. The rows and columns are dependent.

Next, a few specific examples of high-dimensional data are introduced; the roles of covariance matrix estimation in their analysis are emphasized. Descriptions of many other data examples including Web Term-Document Data, Tick-by-Tick Financial Data, Imagery, Hyperspectral Imagery, and many challenges in analyzing high-dimensional data are given in the comprehensive paper by Donoho (2000).

Example 2 (DNA Microarray and Genomics) *Consider the data from an experiment on prostate cancer (Efron, 2010) where there are $p = 6033$ genes and $n = 102$ subjects, of which 50 are healthy controls and 52 are prostate cancer subjects. For each subject, a microarray was run and Y_{ij} is the expression level for person i and gene j. The data matrix is 102×6033 where the rows correspond to subjects and they are independent. The goal is to find a few genes that are involved in prostate cancer. In other words, the researchers hope to find a few key genes that explain the difference between the two groups.*

Estimation of large-scale covariance matrices is a common though often implicit task in DNA, genomics, and transcriptome analysis. Some notable examples are

(a) In clustering genes using data from a microarray experiment, usually a hierarchical tree is constructed to describe the functional grouping of genes. This is based on an estimate of the similarities (distances) between all pairs of expression profiles which are related to the sample correlations. Likewise, the covariance matrix plays a central role in the classification of genes.

(b) In the construction of the so-called relevance networks (Butte et al., 2000) which visually represent the pairwise (in)dependence structure among the p genes. The networks are built by drawing edges between those pairs of genes whose absolute pairwise correlation coefficients exceed a prespecified threshold.

(c) In the construction of gene association networks. These are graphical models which display the conditional (in)dependencies among the genes. The key input to inferring such a network is either the $p \times p$ sample covariance matrix or a sparse precision matrix (see Chapter 5).

In summary, there are numerous bioinformatics tasks that rely on the pairwise sample covariances as the basic ingredients in the analysis. A key question is how to obtain an accurate estimate of the population covariance (precision) matrix based on a dataset with p large and n small?

In the next three examples both the rows and columns are dependent; such data matrices are called *transposable matrices* (Allen and Tibshirani, 2010). Data in neuroimaging, geostatistics, finance, and economics exhibit such dependencies. Conventional multivariate analysis techniques often ignore the dependence in the rows and focus only on accounting for the dependence in the columns.

Example 3 (Neuroimaging and fMRI Data) *Transposable data and the need to account for two-way dependencies in the data matrix arise from various examples in neuroimaging. Functional MRIs (fMRIs) take three-dimensional images of the brain every 2–3 second with high spatial resolution. Each entity in the image, the voxel, measures the blood oxygenation level dependent (BOLD) response, an indirect measure of neuronal activity over time. The fMRI data matrix may have around a hundred thousand voxels (rows) and around a thousand columns or time points. It exhibits strong spatial dependencies among the rows and strong temporal dependencies among the columns (Lindquist, 2008). In spite of this, multivariate analysis techniques are often employed to find regions of interest, or spatially contiguous groups of voxels exhibiting strong activation, and the activation patterns, or the time courses associated with these regions (Lindquist, 2008).*

Example 4 (The Netflix Contest Data) *The Netflix data are prominent examples of sparse, high-dimensional data. It have over* $n = 480,000$ *customers and* $p = 18,000$ *movies, where customers have rated (from 1 to 5) some of their favorite movies but not all. The data are very sparse with a small percentage (about 1%) of the ratings available. The goal is to predict the ratings for unrated movies so that one can have a better recommender system serving the movie interests of different customers. Following the Netflix competition on collaborative filtering, a more recent stream of works has also been focused on exactly reconstructing a low-rank matrix from a small, single incoherent set of observations. A thorough review of the data, methods of analysis, and statistical lessons learned from this contest is given in Feuerverger et al. (2011).*

Example 5 (Asset Returns) *In finance and related areas of business and economics, some important objectives can be reduced to understanding the behavior and dynamics of the asset returns across assets and over time. To set the notation, consider a portfolio of* p *assets held for* T *time periods* $t = 1, \ldots, T$*. For a fixed asset* i*, let* r_{it} *be its log return at time* t*, then* $\{r_{it}\}$ *is a univariate time series of length* T*. The log returns of all the assets in the portfolio, namely* $\{r_{it} : i = 1, \ldots, p; t = 1, \ldots, T\}$ *is a multivariate or* p*-dimensional time series. When viewed as a data matrix its rows and columns are correlated. As an example, Figure 2.1 displays the time series plots of the weekly log returns for three stocks from 2004 to 2010 with* $T = 365$ *observations. Even though the stocks are from three different industry groups, there are considerable similarities in their behavior over time.*

The temporal dependence is usually removed by fitting simple time series models to each row. The returns of an asset and the whole portfolio invariably depend on several economic and financial variables at the macroeconomics and company levels such as the growth rate of GDP, inflation rate, the industry type, company size, and the market value. Techniques from time series and multivariate analysis are central to the study of multiple assets and portfolio returns, and often they involve high-dimensional statistical models with challenging computational issues. Dimension-reduction methods like the PCA and factor analysis are commonly used

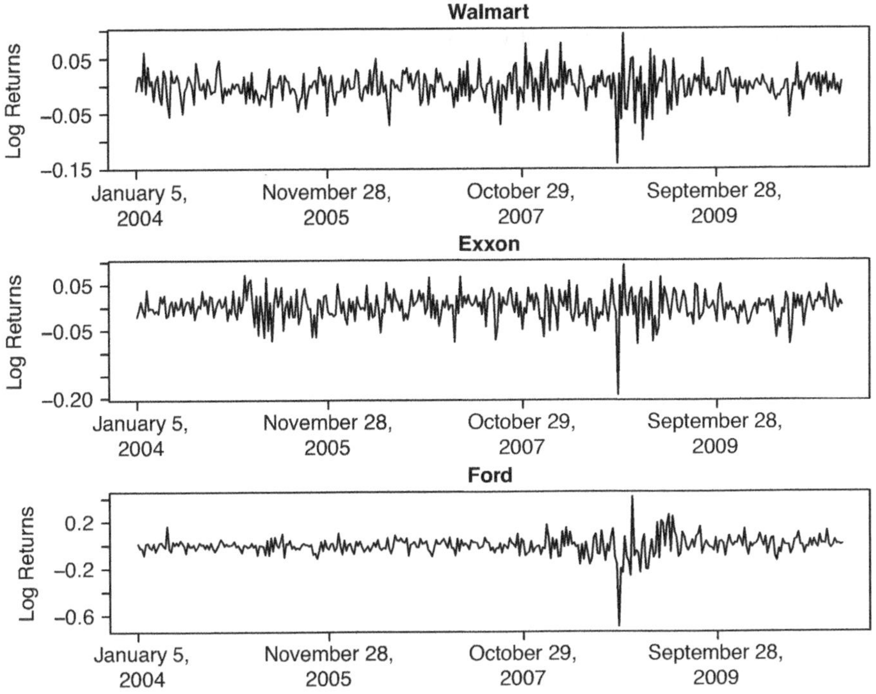

FIGURE 2.1 Time series plots of three stock returns.

in the analysis of financial data. However, the latter has a natural and special role in financial data analysis in the sense that either the subject matter or the statistical analysis of the data can lead to intrinsic and interpretable factors.

The general form of the factor model for a $q < p$, with a slightly different notation than those in Section 1.7, is

$$r_{it} = \alpha_i + \beta_{i1} f_{i1} + \ldots + \beta_{iq} f_{qt} + \varepsilon_{it}, \quad t = 1, \ldots, T; i = 1, \ldots, p, \quad (2.1)$$

where f_{jt}, $j = 1, \ldots, q$ are the q factors, and the β_{ij}'s are the factor loadings. In contrast to the statistical factor models in Section 1.7 where the factors are latent and unobservable, in finance the factors can be observable and hence (2.1) can be viewed as a linear regression model and factor loadings can be estimated using the least-squares method. For example, in macroeconomic factor models *it is common to use macroeconomic variables like growth rate of the GDP, interest rates, and unemployment rate in modeling the behavior of the asset returns. The best known macroeconomic factor model in finance is the single-factor* market model:

$$r_{it} = \alpha_i + \beta_i r_{Mt} + \varepsilon_{it}, \quad i = 1, \ldots, p; t = 1, \ldots, T, \quad (2.2)$$

where r_{it} and r_{Mt} are, respectively, the excess returns of the i-th asset and the market. The S&P 500 *index is usually taken as a proxy for the market (Fama and French, 1992).*

Finally, it is important to note that a high-dimensional data matrix is not necessarily massive or large. It is just the relative size of its dimensions n and p that matters the most as seen in the next example.

Example 6 (Community Abundance Data) *In community ecology, one usually measures the abundance of many different species at a relatively few sites. The dataset in Warton (2008), for example, contains the biomasses of benthic macroinvertebrates. Samples were collected at $n = 12$ different locations, individuals were sorted into species, and the abundance of each species was measured by calculating the total biomass. At each location, $p = 23$ environmental variables were collected. The purpose of the study was to identify whether there is a relationship between community macrobenthic abundance and environmental variables. The key feature of this dataset which is of interest here is that the number of variables $p = 23$ is larger than the number of observations. There are many 0's in the 12×23 data matrix, because not every species is observed in every site. Note that despite the small size of this data matrix, it shares some features of the Netflix data.*

2.2 SHRINKING THE SAMPLE COVARIANCE MATRIX

In this section, some properties and shortcomings of the sample covariance matrix in the high-dimensional situations are reviewed and the need for regularization is pointed out.

The most commonly used summary statistics for a multivariate data matrix $\boldsymbol{Y}$ is its $p \times p$ sample covariance matrix given by

$$\mathbf{S} = \frac{1}{n}\sum_{i=1}^{n}(\boldsymbol{Y}_i - \bar{\boldsymbol{Y}})(\boldsymbol{Y}_i - \bar{\boldsymbol{Y}})',$$

where $\bar{\boldsymbol{Y}} = \frac{1}{n}\sum_{i=1}^{n}\boldsymbol{Y}_i$ is the sample mean. In what follows, we assume that $\bar{\boldsymbol{Y}} = 0$ so that the columns of $\boldsymbol{Y}$ are centered and hence $\mathbf{S} = \frac{1}{n}\boldsymbol{Y}\boldsymbol{Y}'$. In classical multivariate statistics, the number of variables p is usually fixed and small, somewhere between 2 and 10, and n can grow large, then $\mathbf{S}$ is a reliable and undisputed estimator of the population covariance matrix (Anderson, 2003).

It was noted by Stein (1956) that the sample covariance matrix performs poorly when p and p/n are large. In particular, in this case the largest (smallest) eigenvalue of $\mathbf{S}$ tends to be larger (smaller) than its population counterpart. This phenomenon can be seen readily in Figure 2.2 where the population covariance matrix is taken to be the identity matrix with all its eigenvalues equal to one (indicated by the straight dotted lines parallel to the horizontal axis). Note that the range of the sample eigenvalues gets larger for larger values of the ratio p/n. This ratio or its limit plays a fundamental role in every aspect of the asymptotic behavior of the sample eigenvalues of $\mathbf{S}$.

Next, improved estimators of the sample covariance matrix are introduced by shrinking the eigenvalues of $\mathbf{S}$ toward a central value. The amount of shrinkage is

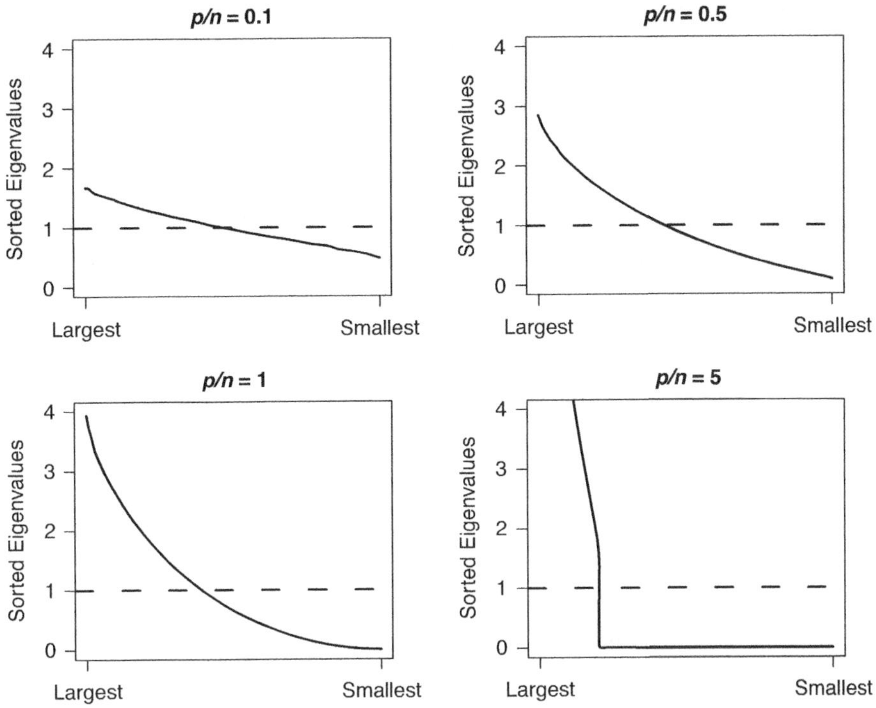

FIGURE 2.2 Plots of the true eigenvalues of the covariance matrix versus sample eigenvalues of the sample covariance matrix. The solid line represents the distribution of the eigenvalues of the sample covariance matrix sorted from largest to smallest, then plotted against their rank. The four plots correspond to the four indicated values of p/n.

determined by minimizing certain risk functions. The two common loss functions with the corresponding risk functions for an arbitrary estimator $\widehat{\boldsymbol{\Sigma}} = \widehat{\boldsymbol{\Sigma}}(\mathbf{S})$ are

$$L_1(\widehat{\boldsymbol{\Sigma}}, \boldsymbol{\Sigma}) = \text{tr}\left(\widehat{\boldsymbol{\Sigma}}\boldsymbol{\Sigma}^{-1}\right) - \log|\widehat{\boldsymbol{\Sigma}}\boldsymbol{\Sigma}^{-1}| - p,$$
$$L_2(\widehat{\boldsymbol{\Sigma}}, \boldsymbol{\Sigma}) = \text{tr}\left(\widehat{\boldsymbol{\Sigma}}\boldsymbol{\Sigma}^{-1} - I\right)^2,$$

and

$$R_i(\widehat{\boldsymbol{\Sigma}}, \boldsymbol{\Sigma}) = E_{\boldsymbol{\Sigma}} L_i(\widehat{\boldsymbol{\Sigma}}, \boldsymbol{\Sigma}), i = 1, 2.$$

An estimator $\widehat{\boldsymbol{\Sigma}}$ is considered better than the sample covariance matrix **S**, if its risk function is smaller than that of **S**. The loss function L_1 was advocated by Stein (1956) and is usually called the entropy loss or the Kullback–Liebler divergence of two multivariate normal densities corresponding to the two covariance matrices. The second, called the quadratic loss function is essentially the Euclidean or the Frobenius norm of its matrix argument which involves squaring the difference between aspects of the estimator and the target. Consequently, it penalizes overestimates more than underestimates, and "smaller" estimates are more favored under L_2 than under L_1.

For example, among all scalar multiples $a\mathbf{S}$, $a > 0$, it is known (Haff, 1980) that $\mathbf{S}$ is optimal under L_1, while the smaller estimator $n\mathbf{S}/(n+p+1)$ is optimal under L_2.

Having observed that the sample covariance matrix systematically distorts the eigenstructure of $\boldsymbol{\Sigma}$, particularly when p/n is large, Stein (1956) considered shirnkage estimators of the form

$$\widehat{\boldsymbol{\Sigma}} = \widehat{\boldsymbol{\Sigma}}(\mathbf{S}) = P\Phi(\lambda)P',$$

where $\lambda = (\lambda_1, \ldots, \lambda_p)'$, $\lambda_1 > \ldots > \lambda_p > 0$ are the ordered eigenvalues of $\mathbf{S}$, P is the orthogonal matrix whose jth column is the corresponding normalized eigenvector, and $\Phi(\lambda) = \mathrm{diag}(\varphi_1, \ldots, \varphi_p)$ is a diagonal matrix where $\varphi_j = \varphi_j(\lambda)$ estimates the jth largest eigenvalue of $\boldsymbol{\Sigma}$. For example, the choice of $\varphi_j = \lambda_j$ corresponds to the usual unbiased estimator $\mathbf{S}$, where it is known that λ_1 and λ_p have upward and downward biases, respectively. Stein's shrinkage method chooses $\Phi(\lambda)$ so as to counteract the biases of the eigenvalues of $\mathbf{S}$ by shrinking them toward a central value. For the L_1 risk, the modified estimators of the eigenvalues of $\boldsymbol{\Sigma}$ for $j = 1, \ldots, p$, are of the form

$$\varphi_j = \lambda_j \cdot n \left(n - p + 1 + 2\,\lambda_j \sum_{i \neq j} \frac{1}{\lambda_j - \lambda_i} \right)^{-1}. \tag{2.3}$$

Note that the φ_j differs the most from λ_j when some or all of the λ_j's are nearly equal and n/p is not small.

Unfortunately, in (2.3) some of the φ_j's could be negative and may not be monotone. Stein suggested an isotonizing procedure to force the modified estimators to satisfy the above constraints. The procedure is explained in more details in Lin and Perlman (1985); through extensive simulation studies they show that the above shrinkage estimator has significant improvement in risk over the sample covariance matrix. It performs the best when the eigenvalues of the population covariance matrix are nearly equal or form clusters within each of which the eigenvalues are nearly equal. They have also applied the idea of Stein shrinkage estimators to the sample correlation matrix by shrinking the Fisher z-transform of the individual correlation coefficients (and the logarithm of the variances) toward a common target value.

A simpler and larger class of well-conditioned shrinkage estimators, based on linear combinations of the sample covariance matrix and the identity matrix, proposed by Ledoit and Wolf (2004), is of the form

$$\widehat{\boldsymbol{\Sigma}}_{\mathrm{LW}} = \alpha_1 I + \alpha_2 \mathbf{S}, \tag{2.4}$$

where α_1 and α_2 are chosen to minimize the risk function corresponding to the Frobenius norm. It subsumes the popular class of empirical Bayes estimators (Haff, 1980; Efron, 2010):

$$\widehat{\boldsymbol{\Sigma}}_{\mathrm{EB}} = \frac{np - 2n - 2}{n^2 p}\alpha_{\mathrm{EB}} I + \frac{n}{n+1}\mathbf{S}, \tag{2.5}$$

where $\alpha_{EB} = [\det(\mathbf{S})]^{1/p}$ and the subscript EB stands for empirical Bayes. These improved estimators of the sample covariance matrix shrink only the eigenvalues of $\mathbf{S}$, and leave intact the eigenvectors. The topic of shrinking the eigenvalues and eigenvectors is discussed in more detail in Chapter 4.

2.3 DISTRIBUTION OF THE SAMPLE EIGENVALUES

In this section, the excessive spread of the sample eigenvalues, seen in Figure 2.2, is explained more rigorously by studying the asymptotic behavior of their empirical distribution.

Consider the *empirical spectral distribution* of $\mathbf{S}_n$, that is, the distribution of its p sample eigenvalues, defined by

$$F_{(n,p)}(t) = \frac{1}{p}\#\{\lambda_i : \lambda_i \le t\},$$

for $t \in R^+$. A milestone result by Marčenko and Pastur (1967) shows that if both the sample size n and data dimension p grow to infinity such that

$$p/n \to c \in (0, \infty),$$

for some positive $c > 0$, then $F_{(n,p)}$ converges to a nonrandom distribution $F(t)$ provided that the entries of $\boldsymbol{Y}_i$ are i.i.d. with mean 0 and variance 1. This limiting spectral distribution, named after them as the Marčenko–Pastur distribution of index c, has a density function of the form

$$f(x) = \frac{1}{2\pi c x}\sqrt{(b-x)(x-a)}, \quad a \le x \le b, \tag{2.6}$$

where $a = (1-\sqrt{c})^2$ and $b = (1+\sqrt{c})^2$ define the support of the distribution, it has a point mass $1 - 1/c$ at the origin if $c > 1$. Note that larger values of c lead to wider support intervals for the sample eigenvalues. In particular, for $c = 1$ the support is the whole interval $[0, 4]$ in complete agreement with Figure 2.2.

What is the limiting distribution of the sample eigenvalues when the components of $\boldsymbol{Y}_i$ are dependent with a general covariance matrix $\boldsymbol{\Sigma}$? This question has been addressed by Johnstone (2001) in the Gaussian case, and by other researchers (Bai and Silverstein, 2010) in the more general case when the distribution of $\boldsymbol{Y}$ is assumed to be sub-Gaussian. In this setup, the largest eigenvalue of $\mathbf{S}_n$ follows the Tracey–Widom distribution for large n, p.

Let $R = (r_{ij})$ be the sample correlation matrix corresponding to $\mathbf{S}_n$ and define

$$L_n = \max_{1 \le i < j \le p} \max |r_{ij}|, \tag{2.7}$$

as the largest magnitude of the off-diagonal entries of R. This statistic is of interest when testing the independence or certain correlation structure in the data, and in the literature of *compressed sensing*, see Example 16 and Donoho (2006), where it is referred to as the *coherence* of the data matrix $\boldsymbol{Y}$. In both applications, it is of interest to know the limiting distribution of L_n when both n, p are large and comparable. Its limiting distribution when the entries of the rows of $\boldsymbol{Y}$ are correlated has been studied by Cai and Jiang (2011) for ultra high-dimensional data or when $p = e^{n^\eta}$ for some $\eta > 0$. Under some regularity conditions on $\boldsymbol{Y}$, the limiting law of the coherence is the extremal distribution of Type I or the Gumbel distribution.

2.4 REGULARIZING COVARIANCES LIKE A MEAN

There is a growing tendency to embed covariance estimation problems within the framework of (nonparametric) mean estimation (Bickel and Levina, 2008a; Cai and Liu, 2011). In this section, we review estimating the mean of a normal random vector using shrinkage (Stein, 1956) and hard-thresholding (Donoho and Johnstone, 1994), respectively. The implications are important for covariance regularization.

The magical interaction between *high-dimensionality* and *sparsity* is explained next using Stein's shrinkage idea when estimating the mean of a homoscedastic normal random vector.

Example 7 (Stein's Phenomenon, Sparsity, and High-Dimensionality) *Consider a p-dimensional random vector* $\boldsymbol{Y} \sim N_p(\boldsymbol{\mu}, I)$ *where* $\boldsymbol{\mu} \neq 0$ *and the goal is to estimate the mean vector using the single* ($n = 1$) *data vector* $\boldsymbol{Y}$. *The risk of any estimator* $\tilde{\boldsymbol{\mu}}$ *is measured using the squared error loss*

$$R(\tilde{\boldsymbol{\mu}}) = E||\tilde{\boldsymbol{\mu}} - \boldsymbol{\mu}||^2.$$

Note that the risk of the maximum likelihood estimator (MLE) $\widehat{\boldsymbol{\mu}} = \boldsymbol{Y}$, *computed using the fact that var*$(Y_i) = 1$ *is*

$$R(\widehat{\boldsymbol{\mu}}) = \sum_{i=1}^{p} E(Y_i - \mu_i)^2 = p\,var(Y_1) = p.$$

Now, consider shrinking the MLE toward the zero (the most sparse) vector. For this, we find β_0, *the minimizer of* $R(\boldsymbol{\mu}_\beta)$ *where* $\boldsymbol{\mu}_\beta = \beta \boldsymbol{Y}$. *Using basic calculus, it follows that*

$$\beta_0 = (1 + p/||\boldsymbol{\mu}||^2)^{-1},$$

with its risk satisfying

$$(1 + p/||\boldsymbol{\mu}||^2)^{-1} p = R(\boldsymbol{\mu}_{\beta_0}) < R(\widehat{\boldsymbol{\mu}}) = p.$$

The difference between the two risks is

$$R(\widehat{\mu}) - R(\mu_{\beta_0}) = \frac{p^2}{p + ||\mu||^2}, \tag{2.8}$$

which gets larger for larger p (higher-dimensional data) and for smaller $||\mu||^2$ *(sparser parameter vector). Indeed, this useful interplay between sparsity and high-dimensionality was the first indication that in higher dimensions it is much easier to beat the standard MLE in the presence of parameter sparsity.*

Unfortunately, μ_{β_0} is not really an *estimator* since it depends on the unknown $||\mu||^2$. A *reasonable* estimator of the shrinkage parameter β_0 can be obtained by replacing $||\mu||^2$ by its unbiased estimator $\sum_{i=1}^{p} Y_i^2 - p$, which leads to the estimator $\widehat{\mu}_0 = (1 - \frac{p}{||Y||^2})Y$. An even better estimator is obtained by setting $\beta_0 = (1 - \frac{p-2}{||Y||^2})$ which leads to the well-known James–Stein estimator of the mean of a high-dimensional normal random vector (Efron, 2010).

Surprisingly, the difference in (2.8) depends only on the dimension p of the parameter vector and its norm $||\mu||^2$, but not on the whole p-dimensional unknown vector μ. A similar simplification occurs for the shrinkage covariance estimator in (2.4) where the optimal coefficients α_1 and α_2 depend only on four simple functionals of Σ (see the Ledoit–Wolf estimator in Chapter 4).

The idea of *hard-thresholding* introduced by Donoho and Johnstone (1994) for estimating sparse normal mean vectors has enjoyed considerable success in nonparametric function estimation in terms of asymptotic optimality and computational simplicity. Here, we show its potential role in high-dimensional covariance estimation.

Example 8 (Thresholding the Mean Vector) *Consider an idealized Gaussian sequence model (Johnstone, 2011) in which*

$$Y_i = \mu_i + \sigma_i Z_i, \quad Z_i \sim i.i.d.N(0,1), 1 \le i \le p, \tag{2.9}$$

and the goal is to estimate a sparse mean vector μ. *If the noise levels* σ_i*'s are bounded, say by b, then the universal thresholding rule*

$$\widehat{\mu}_i = Y_i I(|Y_i| \ge b\sqrt{2\log p}),$$

performs well asymptotically over the ℓ_q *ball:*

$$B_q(s_0) = \left\{ \mu \in R^p : \sum_{j=1}^{p} |\mu_j|^q \le s_0 \right\}. \tag{2.10}$$

In particular, $B_q(s_0)$ *for* $q = 0$ *is a set of sparse vectors with at most* s_0 *nonzero elements. The assumption that* σ_i*'s are bounded is crucial, because then the universal thresholding rule simply treats the heteroscedastic problem (2.9) as a homoscedastic one with all noise levels* $\sigma_i = b$. *It is intuitively clear that this method does not perform well when the range of* σ_i *is large. If the noise level* σ_i*'s are known or can be estimated reasonably well, then a good estimator of the mean vector is the* adaptive thresholding estimator

$$\widehat{\mu}_i = Y_i I\left\{|Y_i| \geq \sigma_i\sqrt{2\log p}\right\}. \tag{2.11}$$

It is certainly tempting to apply the idea of hard-thresholding to the sparse covariance estimation problem with the entries of the sample covariance matrix as the "observations". However, since the entries of the sample covariance matrix have different variances one has to treat the problem as a heteroscedastic regression. In fact, under some mild moment conditions on $\boldsymbol{Y}$ and applying the central limit theorem to the individual entries of the sample covariance matrix, it follows that for n large

$$\frac{1}{n}\sum_{k=1}^{n}(Y_{ki}-\mu_i)(Y_{kj}-\mu_j) \approx \sigma_{ij} + \sqrt{\frac{\theta_{ij}}{n}}Z_{ij}, \quad 1 \leq i, j \leq p, \tag{2.12}$$

where Z_{ij}'s are asymptotically independent standard normal random variables, and $\theta_{ij} = \text{var}[(Y_i - \mu_i)(Y_j - \mu_j)]$ captures the variability of the entries of $\mathbf{S}$. This approximate connection between covariance matrix and mean vector estimation will be used in Section 6.3 to introduce adaptive hard-thresholding of covariance matrices. This is yet another instance where a regularization technique for sparse estimation of the mean vector is useful for sparse covariance estimation.

2.5 THE LASSO REGRESSION

By now, we have seen enough examples showing that the goal of reducing sparse covariance estimation to that of a sequence of regression problems is achievable. Its implementation, however, requires a working knowledge of the Lasso and penalized least-squares regression. In the next few sections, continuing from Section 1.2 we review some additional properties and extensions of the Lasso regression.

The Lasso estimate of the regression coefficients is obtained by minimizing the sum of squares of residuals subject to a constraint on the sum of absolute values of the coefficients:

$$\text{minimize } \frac{1}{2}\|\boldsymbol{Y} - X\beta\|^2 \quad \text{subject to: } \sum_j |\beta_j| \leq c, \tag{2.13}$$

where c is the tuning parameter. It is evident that choosing c sufficiently small will shrink most of the coefficients to zero. The Lagrangian form of (2.13) is

$$Q(\beta) = \frac{1}{2}\|\boldsymbol{Y} - X\beta\|^2 + \lambda\sum_j |\beta_j|, \tag{2.14}$$

where λ is a penalty controlling the sparsity of the model. Its solution turns out to be nonlinear in the responses Y_i's due to the form of the penalty.

Fundamental to understanding the Lasso procedure is the soft-thresholding operator. In Chapter 1, it was noted that solving the Lasso problem (1.7) or (2.14) for $n = p = 1$, a generic observation y and $X = 1$, leads to minimizing the simple objective function, $Q(\beta) = \frac{1}{2}(y - \beta)^2 + \lambda|\beta|$, with the derivative:

$$Q'(\beta) = -y + \beta + \lambda \cdot \text{sign}(\beta) = 0. \tag{2.15}$$

The explicit form of its solution in terms of y, λ, is found by rearranging (2.15) as

$$y = \beta + \lambda \cdot \text{sign}(\beta) = |\beta| \cdot \text{sign}(\beta) + \lambda \cdot \text{sign}(\beta) = (|\beta| + \lambda) \cdot \text{sign}(\beta).$$

Since $|\beta| + \lambda > 0$, it follows that the Lasso solution and y must have the same sign, i.e.,

$$\text{sign}(\beta) = \text{sign}(y). \tag{2.16}$$

Once again, rewrite (2.15) as

$$\beta = y - \lambda \cdot \text{sign}(y) = |y|\text{sign}(y) - \lambda \cdot \text{sign}(y) = \text{sign}(y) \cdot (|y| - \lambda),$$

it follows that the Lasso solution for a fixed penalty λ is

$$\widehat{\beta}(\lambda) = \text{sign}(y) \cdot (|y| - \lambda)_+, \tag{2.17}$$

where $(x)_+ = x$, if $x > 0$, and 0 otherwise. The positive truncation function $(x)_+$ is used here to ensure the condition in (2.16). The Lasso solution as a function of λ in (2.17) is called the *soft-thresholding operator*.

Next, it will be seen that solutions of more general Lasso regression problems can be written in terms of the soft-thresholding operator.

Example 9 (The One-Sample Problem) *For n observations $Y_1, \ldots, Y_n$, $p = 1$ and $X = 1$, $Q(\beta)$ in (2.13) reduces to $\frac{1}{2}\sum_{i=1}^{n}(Y_i - \beta)^2 + \lambda|\beta|$. It follows from the above arguments that the Lasso estimate of β is given by*

$$S(\bar{Y}, \lambda) = \text{sign}(\bar{Y})(|\bar{Y}| - \lambda)_+, \tag{2.18}$$

where $\bar{Y}$ is the sample mean of the n observations.

Example 1 provides another interesting case of using the soft-thresholding operator in estimating the slope in a simple linear regression with no intercept.

Example 10 (Regression with Orthogonal Design Matrices) *Consider a vector of n observations $\mathbf{Y} = (Y_1, \ldots, Y_n)'$ and an $n \times p$ design matrix $X_{n\times p} = (X_{\cdot 1}, \ldots, X_{\cdot p})$*

such that $X'X = I_p$. Here, $X_{\cdot j}, 1 \leq j \leq p$ is an n-vector of the values of the jth predictor variable. The normal equations for the Lasso solution from (2.14) given by

$$\frac{\partial Q(\beta)}{\partial \beta} = -X'Y + \beta + \lambda \cdot \text{sign}(\beta) = 0,$$

are equivalent to those for p separate simple Lasso regressions. Its componentwise solutions are

$$\widehat{\beta}_i(\lambda) = \text{sign}(X'_{\cdot i}Y)(X'_{\cdot i}Y - \lambda)_+ = S(X'_{\cdot i}Y, \lambda),$$

and in vector form the solution can be written as

$$\widehat{\beta}(\lambda) = \text{sign}(X'Y)(|X'Y| - \lambda)_+ = S(|\widehat{\beta}|, \lambda), \tag{2.19}$$

where

$$\widehat{\beta} = X'Y = \left(X'_1 Y, \ldots, X'_p Y\right)',$$

is the familiar least-squares estimate of the vector of regression coefficients.

How does one solve the normal equations for Lasso regression when its design matrix is not orthonormal?

From (2.19) it is evident that, for a fixed penalty λ, the Lasso regression is more likely to force to zero the coefficients of those predictors which are less correlated with the response. Furthermore, the Lasso solutions given in Example 10 are piecewise linear in λ due to piecewise linearity of the soft-thresholding operator noted earlier in Section 2. These observations are the source of a number of insightful and innovative approaches to computing the Lasso solution which rely on the correlations among the response and covariates. These include

1. The *least angle regression* (LAR) algorithm, due to Efron et al. (2004), is closely related to the forward stepwise regression. Recall that the latter builds a model sequentially, adding one variable at a time. LAR uses a similar strategy, but it enters first the variable most correlated with the response. Rather than fit this variable completely, LAR moves the coefficient of this variable continuously toward its least-squares value. It computes the whole solution path of the Lasso regression within $O(n^2 p)$ operations, which is a truly remarkable computational achievement.
2. The *sure independence screening* (SIS), due to Fan and Lv (2008), is based on correlation learning, in the sense that it starts by filtering out the covariates that have weak correlation with the response.

We present the *coordinate descent algorithm* which reduces solving a general Lasso regression problem to that of a sequence of one-dimensional (simple) Lasso regression problems where the soft-thresholding operator can be employed.

Example 11 (The Coordinate Descent Algorithm) *For a general design matrix* $\boldsymbol{X}$*, there is no closed form for the Lasso solution and the computational details of the Lasso procedure are more involved. A fast method to solve the general Lasso regression problem is the* coordinate descent algorithm *which minimizes (1) over one* β_j *at a time with the others kept fixed. It then cycles through all the parameters until convergence (see Friedman et al., 2008). More precisely, suppose all the values of* β_k *for* $k \neq j$ *are held fixed at their current value* $\tilde{\beta}_k$*, so that* $Q(\tilde{\beta})$ *can be written as*

$$Q(\tilde{\beta}) = \frac{1}{2}\sum_{i=1}^{n}\left(Y_i - \sum_{k\neq j} x_{ik}\tilde{\beta}_k - x_{ij}\beta_j\right)^2 + \lambda\sum_{k\neq j}|\tilde{\beta}_k| + \lambda|\beta_j|, \tag{2.20}$$

and viewed as a function of β_j *alone. Minimizing* $Q(\tilde{\beta})$ *with respect to* β_j *amounts to solving a univariate Lasso problem as in (1.12) with no intercept where the responses are* $r_i^{(j)} = Y_i - \sum\limits_{k\neq j} x_{ik}\tilde{\beta}_k$*. Hence, from (1.12) one obtains the explicit solution*

$$\tilde{\beta}_j \leftarrow S\left(\sum_{i=1}^{n} x_{ij} r_i^{(j)}, \lambda\right).$$

Now, the algorithm cycles through all β_j, $j = 1, 2, \ldots, p, 1, 2, \ldots, p, \ldots$ *until* $\tilde{\beta}$ *converges. Thus, it is instructive to think of the coordinate descent algorithm as a way of doing iterative componentwise (one-dimensional) soft-thresholding.*

In general, the characterization of the Lasso solution $\widehat{\beta}$ of (2.14) is derived using the Karush–Kuhn–Tucker (KKT) conditions for the minimizer of convex functions. The result is given in the following lemma. For a proof and more information on the convexity and KKT conditions, see Bühlmann and van de Geer (2011, Section 2.5).

Lemma 2 (Characterization of the Lasso Solution) *Let the gradient of* $\frac{1}{2}||\boldsymbol{Y} - X\beta||^2$ *be denoted by the vector* $G(\beta) = -X'(\boldsymbol{Y} - X\beta)$ *and* $G_j(\cdot)$ *be its* j*th entry. Then, a necessary and sufficient condition for* $\widehat{\beta}$ *to be the minimizer of (2.14) is that*

$$G_j(\widehat{\beta}) = -sign(\widehat{\beta}_j)\lambda \quad if \quad \widehat{\beta}_j \neq 0,$$

$$|G_j(\widehat{\beta})| \leq \lambda \quad if \quad \widehat{\beta}_j = 0,$$

for $j = 1, \ldots, p$.

This simple characterization of the Lasso solution will be used when dealing with the general Lasso regression.

Asymptotic properties of the Lasso estimates for linear regression models, their performance from the perspectives of sparsity, variable selection, prediction, consistency, and efficiency are reviewed briefly in the next two sections, see Bühlmann and van de Geer (2011) for the most comprehensive and up-to-date coverage.

2.6 LASSO: VARIABLE SELECTION AND PREDICTION

For linear regression models, the performance of the Lasso method as a variable selection tool and its impact on prediction are reviewed briefly in this section.

In a linear regression model with no intercept and p covariates, let β^0 be the vector of true parameter values. Then, the set of integers

$$S_0 = \{j : \beta_j^0 \neq 0, j = 1, \ldots, p\}$$

representing the indices of the true nonzero regression coefficients is called the *active set* and $s_0 = \text{card}(S_0) = |S_0|$ is the *sparsity index* of β^0 or the linear model. How good is the Lasso method in selecting the active set from the data?

For a fixed and reasonably large value of the tuning parameter λ the Lasso estimator of some coefficients are exactly zero, then it is plausible to scan the Lasso estimates for larger coefficients in the search for identifying the active set. More precisely, one may start with

$$\hat{S}(\lambda) = \{j : \hat{\beta}_j(\lambda) \neq 0, j = 1, \ldots, p\}, \tag{2.21}$$

and look for conditions on the design matrix $\boldsymbol{X}$ which guarantee Lasso detecting correctly all nonzero coefficients. This turns out to be a difficult and an ambitious task, however, a less ambitious and relevant goal is to consider the task of identifying those covariates where the magnitude of their coefficients is bigger than a certain threshold b. More precisely, for some $b > 0$, consider the b-relevant set of coefficients

$$S_b = \{j : |\beta_j^0| \geq b, j = 1, \ldots, p\}. \tag{2.22}$$

Then, under a *compatibility (restricted eigenvalue or near-orthogonality) condition* on the design matrix $\boldsymbol{X}$ (Bühlmann and van de Geer, 2011), it can be shown that for any $b > 0$, as $n \to \infty$,

$$P[\hat{S}(\lambda) \supset S_b] \to 1. \tag{2.23}$$

This property of Lasso called *variable screening* guarantees that under some conditions, with high probability, the Lasso estimated model includes all the relevant covariates or those with *big* coefficients. The variable screening property of Lasso establishes its great potential in achieving sparsity. A simple way to see this is through another important property of Lasso: *every Lasso estimated model has cardinality smaller or equal to* $\min(n, p)$. This useful fact is a consequence of the analysis leading to the LARS algorithm (Efron et al., 2004). In high-dimensional data where $p \gg n$ and hence $\min(n, p) = n$, Lasso achieves a large reduction in dimensionality reducing the number of covariates from p to n.

For a given Lasso estimator $\hat{\beta}(\lambda)$ of the regression coefficients, the corresponding Lasso predictor of the response vector $\boldsymbol{Y}$ is $\boldsymbol{X}\hat{\beta}(\lambda)$. How good is this predictor compared to its least-square counterpart? Surprisingly, the answer is that the Lasso prediction error is of the same order of magnitude as the prediction error one would have if one knew a priori the relevant variables.

For the normal linear models with p covariates, it follows from the standard least-squares theory that

$$E||\mathrm{X}(\hat{\beta} - \beta^0)||^2 = \sigma^2 p,$$

where $\hat{\beta}$ is the least-squares estimate of β. For the Lasso predictor corresponding to a tuning parameter λ of order $\sqrt{\log p/n}$, it can be shown that the prediction error satisfies a so-called *oracle inequality*:

$$E||\boldsymbol{X}[\hat{\beta}(\lambda) - \beta^0]||^2 = O\left(\frac{s_0 \log p}{\phi^2}\right), \tag{2.24}$$

where ϕ^2 is a constant measuring the degree of compatibility between the design matrix $\boldsymbol{X}$ and the ℓ_1 norm of the nonzero regression coefficients. If s_0 is comparable to p, then comparing the prediction errors of Lasso and LS it follows that up to the $\log p$ term and the compatibility constant ϕ^2, the two mean-squared prediction errors are of the same order. This means that the Lasso method acts as if one knew a priori which of the covariates are relevant and then uses the ordinary least-square estimate of the true, relevant s_0 variables only. The rate in (2.24) is optimal, up to the factor $\log p$ and the inverse compatibility constant $1/\phi^2$. The additional $\log p$ factor can be interpreted as the price one must pay for not knowing the active set S_0.

2.7 LASSO: DEGREES OF FREEDOM AND BIC

Choosing the tuning parameter in Lasso and other regularization methods for high-dimensional data is one of the most critical issues. Larger values of λ are known to kill off too many entries and introduce too much bias, whereas smaller values keep too many entries of β leading to more complex models with too much variance. In such situations, it is common to seek a kind of bias–variance trade-offs using various forms of cross-validation and information criteria such as AIC and BIC.

The key conceptual tool in choosing the Lasso tuning parameter relates to the notion of degrees of freedom in linear models. Fortunately, for the Lasso regression it is known (Zou et al., 2007) that $|\hat{S}(\lambda)|$ or the number of nonzero entries in (2.21) provides an unbiased estimate of the degrees of freedom of the Lasso fit, that is,

$$\hat{df}(\lambda) = |\hat{S}(\lambda)|.$$

Now, for example, the BIC can be defined and used to select the optimal number of nonzero coefficients. More precisely, one chooses the tuning parameter λ by minimizing the BIC criterion:

$$\hat{\lambda}_{\text{BIC}} = \underset{\lambda}{\text{argmin}} \left\{ ||\boldsymbol{Y} - \hat{\boldsymbol{Y}}||^2 + \hat{df}(\lambda) \log n \right\}, \tag{2.25}$$

where $\hat{\boldsymbol{Y}}$ is the vector of fitted values for the response vector $\boldsymbol{Y} = (Y_1, \ldots, Y_n)'$.

Since the solution path of $\hat{\beta}(\lambda)$ is piecewise linear in λ, the minimizer in (2.25) can be found by evaluating the relevant function at n points.

2.8 SOME ALTERNATIVES TO THE LASSO PENALTY

In this section, several alternative penalty functions are introduced, they are designed to remedy some of the drawbacks of the Lasso penalty. The Lasso penalty is popular because of its convexity, but is known to produce biased estimates of the regression coefficients due to the linear increase of the penalty function (Fan and Li, 2001). In other words, it shrinks unnecessarily even the largest regression coefficient.

Fortunately, the optimization problems for most of the alternative methods are that of minimizing a generic function of the form:

$$Q(\beta) = \frac{1}{2}||\boldsymbol{Y} - \boldsymbol{X}\beta||^2 + \sum_{j=1}^{p} p_{\lambda_j}(|\beta_j|), \tag{2.26}$$

where $p_{\lambda_j}(\cdot)$ is a sparsity-inducing penalty function with its own tuning parameter. In what follows, we provide several important examples of the penalty functions with desirable properties (see Example 14).

Example 12 (The Adaptive Lasso) *The adaptive Lasso was introduced by Zou (2006) to correct Lasso's overestimation or bias problem. It replaces the ℓ_1 penalty in Lasso by a weighted version where the weights depend on the data or some initial estimates of the parameters. More precisely, the objective function for adaptive Lasso is*

$$Q(\beta) = \frac{1}{2}\|\boldsymbol{Y} - \boldsymbol{X}\beta\|^2 + \lambda \sum_{j} \frac{|\beta_j|}{|\hat{\beta}_j|}, \tag{2.27}$$

where $\hat{\beta}_j$ is the least-squares estimate of the jth component of the vector of regression parameters. An adaptive Lasso solution is denoted by the vector $\hat{\beta}_a(\lambda)$ where the subscript a stands for adaptive.

The adaptive Lasso solution has two desirable properties: (i) when the LS estimator is small or $\hat{\beta}_j = 0$, then the adaptive Lasso estimator is also small or $\hat{\beta}_{a,j}(\lambda) = 0$ and (ii) when $\hat{\beta}_j$ is relatively large, then little shrinkage is done on the β_j so that the estimator has less bias. Both of these properties are immediate from the form of the weighted penalty. Surprisingly, for orthogonal design matrices the adaptive Lasso has an explicit solution of the form

$$\hat{\beta}_{a,j}(\lambda) = \text{sign}(\hat{\beta}_j)\left(|\hat{\beta}_j| - \frac{\lambda}{|\hat{\beta}_j|}\right)_+, j = 1, \ldots, p \tag{2.28}$$

which is again an estimator of the soft-thresholding type. The adaptive thresholding function is given by

$$S_a(x, \lambda) = \text{sign}(x)\left(|x| - \frac{\lambda}{|x|}\right)_+. \tag{2.29}$$

Example 13 (Elastic Net) *Another limitation of Lasso pointed out by Zou and Hastie (2005) is that in the high-dimensional data situations, the number of variables selected by the Lasso is limited by* $\min(n, p)$. *The elastic net generalizes the Lasso to overcome this drawback, while enjoying its other favorable properties. For nonnegative* λ_1 *and* λ_2, *the objective function for elastic net is given by*

$$Q(\beta) = (1 + \lambda_2)\left\{||\boldsymbol{Y} - \boldsymbol{X}\beta||^2 + \lambda_1 \sum_{i=1}^{p} |\beta_i| + \lambda_2 \sum_{i=1}^{p} |\beta_i|^2\right\}, \tag{2.30}$$

where the penalty is a linear combination of the ridge and Lasso penalties. Note that the Lasso (ridge) is a special case of the elastic net when $\lambda_2 = 0 \quad (\lambda_1 = 0)$. *Given a fixed* λ_2, *the LARS–EN algorithm (Zou and Hastie, 2005) efficiently solves the elastic net problem for all* λ_1 *with the computational cost of a single least-squares fit. When* $p > n$, *and* $\lambda_2 > 0$, *the elastic net can potentially include all variables in the fitted model, and hence remove this particular limitation of the Lasso. As expected, for orthogonal design matrices the elastic net has an explicit solution.*

Example 14 (Smoothly Clipped Absolute Deviation (SCAD)) *In the search for an ideal penalty function, Fan and Li (2001) advocate the use of regularization methods leading to estimators having the following desirable properties:*

1. ***Sparsity:*** *The resulting estimator is a thresholding rule, it sets small estimated coefficients to zero to accomplish variable selection.*
2. ***Unbiasedness:*** *The resulting estimator should have low bias, particularly for larger parameter values.*
3. ***Continuity:*** *The resulting estimator should be continuous in the data to reduce instability in the model prediction.*

To provide some insights into these properties and the kind of restrictions they entail on the penalty functions, we consider the following generic penalized-least squares objective function:

$$Q(\beta) = \frac{1}{2}(y - \beta)^2 + p_\lambda(|\beta|). \tag{2.31}$$

Its first-order derivative with respect to β is given by

$$Q'(\beta) = sign(\beta)\{|\beta| + p'_\lambda(|\beta|)\} - y, \tag{2.32}$$

so that the penalized least-squares estimators will have the sparsity, unbiasedness, and continuity properties (Fan and Li, 2001), respectively, provided that

$$\begin{aligned} &\min_\beta\{|\beta| + p'_\lambda(|\beta|)\} > 0; \\ &p'_\lambda(|\beta|) = 0 \quad \textit{for large} \quad |\beta|; \\ &\operatorname*{argmin}_\beta \left\{|\beta| + p'_\lambda(|\beta|)\right\} = 0. \end{aligned} \tag{2.33}$$

For the commonly used ℓ_q penalty function $p(\beta) = |\beta|^q$, it can be shown that:

(i) The ℓ_1 penalty or Lasso does not have the unbiasedness property,
(ii) For $q > 1$ it does not satisfy the sparsity condition,
(iii) Finally, for $0 \leq q < 1$ the continuity condition is not satisfied.

Thus, the ℓ_q penalty function does not satisfy simultaneously the mathematical conditions for sparsity, unbiasedness, and continuity for any q.

In a sense, the sparsity, unbiasedness, and continuity properties of the penalized least squares force the penalty function to be nondifferentiable at the origin and nonconvex over $(0, \infty)$. Thus, to enhance the desirable properties of the ℓ_1 penalty function, one may consider the SCAD penalty which is a continuous and differentiable function defined by

$$p'_{\lambda,a}(x) = \lambda I(|x| \leq \lambda) + \frac{(a\lambda - |x|)_+}{a - 1} I(|x| > \lambda), \tag{2.34}$$

where $a > 2$ is a constant. It corresponds to a quadratic spline function with knots at λ and $a\lambda$ and the value of $a = 3,7$ is suggested for practical use (Fan and Li, 2001). The objective function with the SCAD penalty is

$$Q(\beta) = \frac{1}{2}\|\boldsymbol{Y} - \boldsymbol{X}\beta\|^2 + \sum_{j=1}^{p} p_{\lambda,a}(\beta_j), \tag{2.35}$$

where $p_{\lambda,a}(x)$ is a penalty function with $p_{\lambda,a}(0) = 0$, nondifferentiable at zero. See Fan and Li (2001) for more details on the properties of the SCAD penalty function and the related penalized least-squares procedure.

Example 15 (The Group Lasso) *Imagine a regression problem with a design matrix X with p variables that can be divided into J groups with p_j variables in each with the corresponding design matrix $\boldsymbol{X}_j$. The group Lasso is concerned with minimizing the objective function*

$$Q(\beta) = \frac{1}{2}||\boldsymbol{Y} - \sum_{j=1}^{J} X_j \beta_j||^2 + \lambda \sum_{j=1}^{J} ||\beta_j||, \tag{2.36}$$

where β_j is the coefficient vector for the jth group, and $||\cdot|| = ||\cdot||_2$ is the Euclidean norm. The estimating equations for $j = 1, \ldots, J$ are

$$\boldsymbol{X}_j' \left(\boldsymbol{Y} - \sum_{i=1}^{J} \boldsymbol{X}_i \beta_i \right) - \lambda \delta_j = 0, \tag{2.37}$$

where $\delta_j = \beta_j / ||\beta_j||$ if $\beta_j \neq 0$, and otherwise δ_j is a vector with $||\delta_j|| \leq 1$. These equations can be solved by blockwise coordinate descent in analogy with the coordinate descent algorithm in Example 11. It suffices to focus on the solution for one block, holding the others fixed. Let $\mathbf{r}_j = \boldsymbol{Y} - \sum_{k \neq j} \boldsymbol{X}_k \hat{\beta}_k$ be the partial residual of the jth group and define $\mathbf{s}_j = \boldsymbol{X}_j' \mathbf{r}_j$. It follows that if $||\mathbf{s}_j|| \leq \lambda$, then $\hat{\beta}_j = 0$, otherwise the solution is given by

$$\hat{\beta}_j = \left(\boldsymbol{X}_j' \boldsymbol{X}_j + \frac{\lambda}{||\hat{\beta}_j||} \mathbf{I} \right)^{-1} \boldsymbol{X}_j' \mathbf{r}_j, \tag{2.38}$$

which is similar to the solution of a ridge regression problem, but with the ridge penalty depending on $||\hat{\beta}_j||$. When the X_j is orthonormal, that is, if $\boldsymbol{X}_j' \boldsymbol{X}_j = \mathbf{I}$, then the solution has the following simple form:

$$\hat{\beta}_j = \left(1 - \frac{\lambda}{||\mathbf{s}_j||} \right)_+ \mathbf{s}_j. \tag{2.39}$$

Example 16 (Compressed Sensing) *The theory of compressed sensing or compressed sampling is a rapidly developing area dealing with efficient data acquisition technique that enables accurate reconstruction of highly undersampled sparse signals (Donoho, 2006). It has a wide range of applications including signal processing, medical imaging, and seismology. The theory is closely related to and provides crucial insights into the Lasso regression with many variables. The key difference is that the number of nonzero entries of β is known in advance, whereas in linear regression models it is not.*

Compressed sensing relies on the following two key assumptions:

1. Sparsity (compressibility) *reflects the fact that a small number n of measurements $\boldsymbol{Y}$ can be represented as a linear transformation of a known matrix* *(dictionary)* $X_{n\times p}$ *with* $n \ll p$, *that is,* $Y = X\beta$, *where* β *is* k*-sparse with* k *nonzero entries.*

2. Incoherence *of the sensing matrix* X *which reflects the requirements that any* k *subset of its* p *columns are as orthogonal (uncorrelated) as they can be.*

The main goal of compressed sensing is to construct measurement matrices X, *with the number of measurements* n *as small as possible relative to* p, *such that any* k*-sparse signal* $\beta \in R^p$, *can be recovered exactly from the linear measurements* $Y = X\beta$ *using a computationally efficient recovery algorithm. In compressed sensing, it is typical that* $p \gg n$. *In fact, it is often desirable to make* p *as large as possible relative to* n.

The problem actually relates to solving the underdetermined system of equations in linear algebra where the method of ℓ_1 minimization provides an effective way of recovering the sparse signal β in the above formulation. However, for this to work well, the measurement matrices X must satisfy certain *near-orthogonality* conditions. Two commonly used conditions are the so-called restricted isometry property (RIP) and mutual incoherence property (MIP). Roughly speaking, the RIP requires subsets of certain cardinality of the columns of X to be close to an orthonormal system, and the MIP requires the pairwise correlations among the column vectors of X to be small (see Candes and Tao, 2007).

It is instructive to motivate RIP by considering the case where the locations of the k nonzero entries of β are known. Then, for any $n \geq k$ it is possible to estimate these nonzero entries using the least-squares method, provided that the submatrix X_k corresponding to the k subcolumns is nonsingular, or better yet well-conditioned. In reality, since the locations of the k nonzero entries are unknown, one must assume that any submatrix of k columns of X must be well-conditioned. An elegant way of handling this is through the RIP (Candes and Tao, 2007): there exists a constant $0 < \delta_k < 1$ such that for any k-sparse $\beta \in R^p$, that is, with $||\beta||_0 \leq k$,

$$(1 - \delta_k)||\beta||^2 \leq ||X\beta||^2 \leq (1 + \delta_k)||\beta||^2. \tag{2.40}$$

In other words, a data matrix X satisfying RIP approximately preserves the lengths of all k-sparse vectors so long as the constant δ_k is not too close to 1. Consequently, the k-sparse vectors cannot be in the nullspace of X so that they will not be sacrificed.

A comprehensive review of some of the alternatives to and improvements of Lasso, its history, and computational developments is given by Tibshirani (2011, Table 1).

PROBLEMS

1. Use computer simulation and produce analogs of the plots in Figure 2.2 for $p/n = 3, 4$.

2. **(a)** Show that the sample eigenvalues are more dispersed around the average than the true ones, that is,

$$E\sum_{1}^{p}(\hat{\lambda}_i - \bar{\lambda})^2 = \sum_{1}^{p}(\lambda_i - \bar{\lambda})^2 + pE||\mathbf{S} - \mathbf{\Sigma}||_F^2,$$

where $\hat{\lambda}_i, \lambda_i, i = 1, \ldots, p$ are the eigenvalues of $\mathbf{S}$ and $\mathbf{\Sigma}$, respectively, and

$$\bar{\lambda} = \frac{1}{p}\sum_{1}^{p}\lambda_i$$

is the average of the true eigenvalues of $\mathbf{\Sigma}$.

(b) Let A be a symmetric matrix and P any orthogonal matrix. It is known that $\mathrm{tr}P'AP = \mathrm{tr}A$ and hence the average of the diagonal entries of the two matrices are the same. Show that *the eigenvalues of A are the most dispersed diagonal elements that can be obtained by rotation* (Ledoit and Wolf, 2004).

3. In the setup of Example 7,

(a) Compute the risk of $\boldsymbol{\mu}_{\beta_0}$.

(b) Show that $\sum_1^p Y_i^2 - p$ is an unbiased estimator of $||\boldsymbol{\mu}||^2$.

(c) Is $\hat{\boldsymbol{\mu}}_0$ an unbiased estimator of $\boldsymbol{\mu}$?

4. In the setup of Example 8,

(a) Is the estimator in (2.11) an unbiased estimator of μ_i?

(b) Compute (approximate) the risk of the estimator in (2.11).

(c) Compute $\theta_{ij} = \mathrm{var}\left[(Y_i - \mu_i)(Y_j - \mu_j)\right]$.

CHAPTER 3

COVARIANCE MATRICES

In this chapter, we provide a potpourri of some basic mathematical and statistical results on covariance matrices which are of interest both in the classical multivariate statistics and in the modern high-dimensional data analysis. Included topics are spectral, Cholesky, and singular value decompositions; structured covariance matrices, principal component and factor analysis; generalized linear models (GLMs); and aspects of Bayesian analysis of covariance matrices.

A covariance matrix carries information about the pairwise dependence among the components of a random vector. It is characterized by the *nonnegative definiteness* which is a constraint on all the entries of the matrix. However, information on the dependence of more than two variables at a time such as regression, graphical models, and principal component analysis invariably requires working with the eigenvalues–eigenvectors (spectral decomposition), inverse, logarithm, square-root, and the Cholesky factor of the covariance matrix. These functionals or factors can lead to statistically interpretable and unconstrained reparameterizations of a covariance matrix which could facilitate the task of modeling high-dimensional covariance matrices.

3.1 DEFINITION AND BASIC PROPERTIES

Let $\boldsymbol{Y} = (Y_1, \ldots, Y_p)'$ be a p-dimensional random vector with the *mean vector*

$$\boldsymbol{\mu} = E(\boldsymbol{Y}) = (\mu_1, \ldots, \mu_p)', \tag{3.1}$$

High-Dimensional Covariance Estimation, First Edition. Mohsen Pourahmadi.

where $\mu_i = E(Y_i)$ is the mean of the ith component. Its *covariance matrix* is defined as the $p \times p$ matrix

$$\Sigma = E(\boldsymbol{Y} - \boldsymbol{\mu})(\boldsymbol{Y} - \boldsymbol{\mu})' = (\sigma_{ij}), \tag{3.2}$$

where $\sigma_{ij} = E(Y_i - \mu_i)(Y_j - \mu_j)$ is, indeed, the covariance between the two random variables Y_i and Y_j for $i \neq j$, and σ_{ii} is simply the variance of the ith component of $\boldsymbol{Y}$. The following basic properties of the mean and covariance of a linear transformation of a random vector which follow from (3.1) and (3.2) are used frequently in the sequel.

Lemma 3 *For $\boldsymbol{Y}$ a $p \times 1$ random vector with covariance matrix cov($\boldsymbol{Y}$), and suitable (nonrandom) matrices A, B, C, and D, we have*

(a) $E(A\boldsymbol{Y} + B) = AE(\boldsymbol{Y}) + B$,
(b) $cov(A\boldsymbol{Y} + B) = Acov(\boldsymbol{Y})A'$,
(c) $cov(A\boldsymbol{Y} + B,\ C\boldsymbol{Y} + D) = Acov(\boldsymbol{Y})C'$.

The result in part (b) suggests a way to construct more elaborate covariance matrices from the simpler ones. For example, starting from a $\boldsymbol{Y}$ with the identity covariance matrix, it follows that AA' is a genuine covariance matrix for any matrix A. In fact, all covariance matrices turn out to have this form. In other words, a matrix $\boldsymbol{\Sigma}$ is a covariance matrix, if and only if it is of the form AA' for some matrix A. This reenforces the view that, just like the variance of a random variable, the covariance matrix of a random vector is "nonnegative" in the sense that for any vector $\mathbf{c} = (c_1, \ldots, c_p)'$ the variance of the linear combination $\mathbf{c}'\boldsymbol{Y}$ is nonnegative. More precisely,

$$\mathbf{c}'\boldsymbol{\Sigma}\mathbf{c} = \sum_{i=1}^{p}\sum_{j=1}^{p} c_i c_j \sigma_{ij} = E\left[\mathbf{c}'(\boldsymbol{Y} - \mu)\right]^2 = \text{var}\left[\mathbf{c}'(\boldsymbol{Y} - \mu)\right] \geq 0. \tag{3.3}$$

Motivated by this, we say a symmetric matrix $\boldsymbol{\Sigma}$ is *nonnegative definite* if it satisfies (3.3). It is said to be *positive definite* if the inequality (3.3) is strict for all nonzero vectors $\mathbf{c}$.

Note that if the equality in (3.3) holds for a nonzero vector $\mathbf{c} = \mathbf{c}_0 \neq 0$, then

$$\mathbf{c}_0'(\boldsymbol{Y} - \mu) = \sum_{i=1}^{p} c_{0,i}(Y_i - \mu_i) = 0, \tag{3.4}$$

so that the random variables $Y_1, \ldots, Y_p$ are *linearly dependent*. To avoid this kind of redundancies, in the sequel we often work with the positive-definite covariance matrices.

Although property (3.3) looks deceptively simple, verifying it is a computationally and/or analytically challenging task. Some consequences of (3.3) or equivalent conditions for the nonnegative definiteness are given next.

Theorem 1 *For a $p \times p$ symmetric matrix $\mathbf{\Sigma} = (\sigma_{ij})$, the following are equivalent:*

(a) *$\mathbf{\Sigma}$ is nonnegative definite.*

(b) *all its* leading principal minors *are nonnegative definite, that is, the $i \times i$ matrices*

$$\mathbf{\Sigma}_{ii} = \begin{pmatrix} \sigma_{11} & \cdots & \sigma_{1i} \\ \vdots & \ddots & \vdots \\ \sigma_{i1} & \cdots & \sigma_{ii} \end{pmatrix}, \; i = 1, \cdots, p,$$

are nonnegative definite.

(c) *all eigenvalues of Σ are nonnegative.*

(d) *there exists a matrix A such that*

$$\mathbf{\Sigma} = AA'. \tag{3.5}$$

(e) *there exists a lower triangular matrix L such that*

$$\mathbf{\Sigma} = LL'. \tag{3.6}$$

(f) *there exist vectors $\boldsymbol{u}_1, \cdots, \boldsymbol{u}_p$ in R^p such that $\sigma_{ij} = \boldsymbol{u}_i'\boldsymbol{u}_j$.*

Proof of the last four parts of the theorem relies on the spectral decomposition, square-root, and the Cholesky decomposition of a symmetric matrix, topics which are discussed later in this chapter. In view of part (f), $\mathbf{\Sigma}$ is also called the *Gram matrix* of the vectors $\boldsymbol{u}_1, \cdots, \boldsymbol{u}_p$. Given a symmetric matrix, a great deal of ingenuity is needed to establish its nonnegative definiteness; see Section 3.3 and Bhatia (2006) for several concrete examples and other concepts closely related to the nonnegative-definite matrices and functions. Golub and Van Loan (1996) is an excellent reference for matrices.

The *(Hadamard) Schur product or entrywise product* of two square matrices $A = (a_{ij})$, $B = (b_{ij})$ of the same size is the matrix $A \circ B = (a_{ij}b_{ij})$. It plays a key role in *tapering* covariance matrices in Chapter 6 due to the following striking and useful property.

Theorem 2 (Schur Theorem) *If A and B are nonnegative definite matrices, then so is $A \circ B$.*

Proof *We give a simple probabilistic proof that illustrates the use of Theorem 1(f) and some other basic facts about covariances. Since A, B are nonnegative definite matrices, there exist independent zero-mean normal random vectors*

$X = (X_1, \cdots, X_p)$ and $Y = (Y_1, \cdots, Y_p)$ with A and B as their covariance matrices, respectively. Defining the random vector $Z = (X_1Y_1, \cdots, X_pY_p)'$, it suffices to show that $cov(Z) = A \circ B$. From the independence of X, Y, it is immediate that

$$cov(Z_i, Z_j) = E(X_iY_iX_jY_j) = E(X_iX_j)E(Y_iY_j) = cov(X_i, X_j)cov(Y_i, Y_j) = a_{ij}b_{ij}.$$

A symmetric matrix $A = (a_{ij})$ is said to be *diagonally dominant* if for $i = 1, \cdots, p$,

$$|a_{ii}| > \sum_{j=1, j \neq i}^{p} |a_{ij}|.$$

The following theorem indicates that diagonally dominant matrices might provide some clues as how to band, taper, or threshold a sample covariance matrix so that the sparse estimator remains positive definite (see Chapter 6).

Theorem 3 (Hadamard) *A diagonally dominant matrix with $a_{ii} > 0, i = 1, \cdots, p$, is positive definite.*

One of the simplest decompositions of a covariance matrix expresses $\mathbf{\Sigma}$ in terms of the standard deviations and correlations of the random variables:

$$\mathbf{\Sigma} = DRD, \tag{3.7}$$

where $D = \text{diag}(\sqrt{\sigma_{11}}, \ldots, \sqrt{\sigma_{pp}})$ is a diagonal matrix with the standard deviations of the variables as its diagonal entries and $R = (r_{ij})$ is the *correlation matrix* of $\boldsymbol{Y}$ where

$$r_{ij} = \text{corr}(Y_i, Y_j) = \frac{\sigma_{ij}}{\sqrt{\sigma_{ii}\sigma_{jj}}}.$$

From Theorem 1(b) it follows that $|\sigma_{ij}| \leq \sqrt{\sigma}_{ii}\sqrt{\sigma}_{jj}$, so that the correlation coefficient between any two random variables is less than 1 in absolute value, that is, $|r_{ij}| \leq 1$. Often, it is useful to think of the matrix R as the covariance matrix of the standardized random variables $(Y_i - \mu_i)/\sqrt{\sigma_{ii}}$, $i = 1, \ldots, p$. One may use the matrix

$$\begin{pmatrix} \sqrt{\sigma_{11}} & \sigma_{12} & \cdots & \sigma_{1p} \\ r_{12} & \ddots & & \vdots \\ \vdots & & \ddots & \vdots \\ r_{1p} & \cdots & \cdots & \sqrt{\sigma_{pp}} \end{pmatrix} \tag{3.8}$$

to summarize (store) the nonredundant entries of $\mathbf{\Sigma}$, D, and R. Note that from the matrix (3.8), one can easily construct the three matrices Σ, D, and R.

While every covariance matrix is *symmetric*, that is, $\sigma_{ij} = \sigma_{ji}$ for all i and j, with nonnegative diagonal entries, not every symmetric matrix with nonnegative diagonal entries is the covariance matrix of a random vector. The first example below illustrates this point, while the second larger matrix indicates the need for studying the deeper properties of a covariance matrix like its spectral and Cholesky decompositions.

Example 17

(a) The 2×2 *symmetric matrix*

$$A = \begin{pmatrix} 1 & 2 \\ 2 & 1 \end{pmatrix},$$

with positive diagonal entries cannot be a covariance matrix, since the ensuing correlation coefficient:

$$r_{12} = \frac{a_{12}}{\sqrt{a_{11}a_{22}}} = 2,$$

would be bigger than one in absolute value.

(b) The 5×5 *matrix*

$$A = \begin{pmatrix} 17.5 & & & & \\ 0.55 & 13.2 & & & \\ 0.55 & 0.61 & 10.6 & & \\ 0.41 & 0.49 & 0.71 & 14.8 & \\ 0.39 & 0.44 & 0.66 & 0.61 & 17.3 \end{pmatrix} \tag{3.9}$$

written compactly as (3.8) passes the first test of being a covariance matrix in the sense that the correlation coefficients are less than one in absolute value. However, information about its eigenvalues is needed to assess if it is a genuine covariance matrix.

3.2 THE SPECTRAL DECOMPOSITION

The spectral or the eigenvalue–eigenvector decomposition of a square matrix is a powerful tool in matrix computation and analysis. For rectangular matrices, a similar role is played by the SVD.

A square matrix A is said to have an *eigenvalue* λ, with the corresponding *eigenvector* $x \neq 0$, if

$$Ax = \lambda x. \tag{3.10}$$

For this to hold for a nonzero x, the eigenvalue λ must satisfy the determinantal equation $|A - \lambda I| = 0$ which is a polynomial equation of degree p with possibly complex-valued roots. In what follows, we normalize x so that it has unit length, namely $x'x = 1$ and denote such a generic normalized eigenvector by $\mathbf{e}$.

Example 18 (Real and Complex Eigenvalues)

(a) *The eigenvalues of the* 2×2 *matrix* $A = \begin{pmatrix} 1 & 4 \\ -1 & 1 \end{pmatrix}$ *satisfy the quadratic equation*

$$|A - \lambda I| = \begin{vmatrix} 1-\lambda & 4 \\ -1 & 1-\lambda \end{vmatrix} = (1-\lambda)^2 + 4 = 0.$$

Its roots $\lambda = 1 \pm 2i$, *with* $i = \sqrt{-1}$ *are complex-valued. It is instructive to note that the matrix A is not symmetric.*

(b) *The eigenvalues of the* 2×2 *symmetric matrix* $A = \begin{pmatrix} 1 & 2 \\ 2 & 1 \end{pmatrix}$ *satisfy the quadratic equation*

$$|A - \lambda I| = \begin{vmatrix} 1-\lambda & 2 \\ 2 & 1-\lambda \end{vmatrix} = (1-\lambda)^2 - 4 = 0.$$

Its two roots $\lambda_1 = 1 + 2 = 3$ *and* $\lambda_2 = 1 - 2 = -1$ *are real-valued and ordered from the largest to the smallest. The eigenvector corresponding to, say,* λ_2 *is computed from the system,*

$$(A - \lambda_2 I)x = \begin{pmatrix} 2 & 2 \\ 2 & 2 \end{pmatrix}\begin{pmatrix} x_1 \\ x_2 \end{pmatrix} = \begin{bmatrix} 2x_1 + 2x_2 \\ 2x_1 + 2x_2 \end{bmatrix} = \begin{bmatrix} 0 \\ 0 \end{bmatrix},$$

which reduces to the single equation $x_1 + x_2 = 0$ *with two unknowns. Thus, it has infinitely many solutions of the form* $x_1 = c$, $x_2 = -c$, *where c is a scalar. Normalizing the eigenvector provides an opportunity to get rid of the c, so the normalized eigenvector corresponding to* λ_2 *is* $\mathbf{e}_2 = (-\frac{1}{\sqrt{2}}, \frac{1}{\sqrt{2}})'$. *Similarly, the normalized eigenvector corresponding to* $\lambda_1 = 3$ *is* $\mathbf{e}_1 = (\frac{1}{\sqrt{2}}, \frac{1}{\sqrt{2}})'$. *Interestingly,* $\mathbf{e}_1$ *and* $\mathbf{e}_2$ *are* orthogonal *to each other, that is,* $\mathbf{e}_1'\mathbf{e}_2 = 0$ *and hence the* 2×2 *matrix*

$$P = (\mathbf{e}_1, \mathbf{e}_2)$$

is an orthogonal matrix, *that is,*

$$PP' = P'P = I.$$

Moreover, routine computation shows that the matrix P diagonalizes A, namely,

$$P'AP = \Lambda = \begin{pmatrix} 3 & 0 \\ 0 & -1 \end{pmatrix},$$

where the diagonal entries of Λ *are precisely the ordered eigenvalues of A.*

(c) *The eigenvalues of the* 2×2 *matrix* $A = \begin{pmatrix} 2 & 1 \\ 1 & 2 \end{pmatrix}$ *satisfy the quadratic equation*

$$|A - \lambda I| = \begin{vmatrix} 2-\lambda & 1 \\ 1 & 2-\lambda \end{vmatrix} = (2-\lambda)^2 - 1 = 0.$$

Its roots $\lambda_1 = 2 + 1 = 3$ *and* $\lambda_2 = 2 - 1 = 1$ *are both positive and their corresponding normalized eigenvectors can be computed as in (b). Is this matrix positive definite?*

(d) *The eigenvalues and the corresponding eigenvectors of the* 5×5 *covariance matrix A in Example 17(b) computed using a computer are*

$$\begin{array}{llrrrrr}
\lambda_1 = 688.58, & \mathbf{e}_1 = & (0.51, & 0.37, & 0.34, & 0.45, & 0.54) \\
\lambda_2 = 201.82, & \mathbf{e}_2 = & (-0.75, & -0.20, & 0.07, & 0.30, & 0.55) \\
\lambda_3 = 103.78, & \mathbf{e}_3 = & (-0.31, & 0.45, & 0.15, & 0.57, & -0.59) \\
\lambda_4 = 84.59, & \mathbf{e}_4 = & (0.28, & -0.77, & 0.01, & 0.55, & -0.20) \\
\lambda_5 = 32.40, & \mathbf{e}_5 = & (0.09, & 0.18, & -0.92, & 0.29, & 0.14)
\end{array}$$

The properties of the eigenvalues and eigenvectors of the 2×2 symmetric matrix A, noted in Example 18(b), hold for any symmetric matrix. For ease of reference, these and other properties of symmetric and nonnegative definite matrices are summarized in the following theorem (proofs and more discussions can be found in Golub and Van Loan, 1996):

Theorem 4 (The Spectral Decomposition) *Let A be a* $p \times p$ *symmetric matrix with p pairs of eigenvalues and eigenvectors*

$$(\lambda_1, \mathbf{e}_1),\ (\lambda_2, \mathbf{e}_2), \cdots, (\lambda_p, \mathbf{e}_p).$$

Then,

(a) *The eigenvalues* $\lambda_1, \ldots, \lambda_p$ *are all real, and can be ordered from the largest to the smallest*

$$\lambda_1 \geq \lambda_2 \geq \ldots \geq \lambda_p.$$

(b) *The normalized eigenvectors* $\mathbf{e}_1, \ldots, \mathbf{e}_p$ *are mutually orthogonal and the matrix*

$$P = (\mathbf{e}_1, \mathbf{e}_2, \ldots, \mathbf{e}_p) \tag{3.11}$$

is an orthogonal matrix, *that is,*

$$PP' = P'P = I. \tag{3.12}$$

(c) *The* spectral decomposition *of A is the expansion*

$$A = \lambda_1 \mathbf{e}_1\mathbf{e}_1' + \lambda_2 \mathbf{e}_2\mathbf{e}_2' + \cdots + \lambda_p \mathbf{e}_p\mathbf{e}_p' = P\Lambda P', \tag{3.13}$$

where P is as above and

$$\Lambda = diag(\lambda_1, \cdots, \lambda_p), \tag{3.14}$$

is a diagonal matrix *with* $\lambda_1, \ldots, \lambda_p$ *as its respective diagonal entries.*

(d) *The matrix A is* nonnegative definite, *if and only if all its eigenvalues are nonnegative.*

Part (d) is particularly useful for examining the nonnegative definiteness of matrices when their eigenvalues are known or can be computed numerically as the following example indicates.

Example 19 (Examining Nonnegative Definiteness via Eigenvalues) *The matrix A in Example 18(b) is not nonnegative definite because it has a negative eigenvalue. However, the matrices in Example 18(c,d) are positive definite since all their eigenvalues are positive.*

It is of interest to show that two matrices that are close in some sense have eigenvalues and eigenvectors that are also close. This statement can be made precise by using some commonly used norms for matrices.

For a matrix $A = (a_{ij})$, its squared Frobenius norm is defined by $\|A\|_F^2 = \text{tr}(A'A) = \sum\sum a_{ij}^2$, which is the sum of squares of all its entries. More generally, for A, a $p \times p$ matrix and $0 \leq r, s \leq \infty$, we define

$$\|A\|_{(r,s)} \quad = \quad \max\{\|Ax\|_s \ ; \ x \in R^p, \ \|x\|_r = 1\}$$

where $||x||_r$ is the rth norm of the vector x. It can be shown that $||A||_{(2,2)}$ is the *operator* or the *spectral* norm of A which is also defined by $||A||^2 = \lambda_{\max}(A'A)$, where $\lambda_{\max}(\Sigma)$ is the largest magnitude of its eigenvalues. When A is a symmetric matrix, then $||A|| = |\lambda_{\max}|$.

The *condition number* of a nonnegative definite matrix is defined as λ_1/λ_p or the ratio of its largest and smallest eigenvalues, it is infinity when the latter is zero.

The following results due to Davis and Kahan (1970) are useful in linking the closeness of the eigenstructures of two matrices to the operator norm of their difference.

Theorem 5 *Let* $\{\lambda_i\}$, $\{\widehat{\lambda}_i\}$, $\{e_i\}$, *and* $\{\widehat{e}_i\}$ *be the eigenvalues and associated eigenvectors of the two* $p \times p$ *matrices* $\mathbf{\Sigma}$, $\widehat{\mathbf{\Sigma}}$, *respectively, where the eigenvalues are in descending order. Then,*

(a) *(Weyl's Theorem):* $|\widehat{\lambda}_i - \lambda_i| \leq ||\widehat{\mathbf{\Sigma}} - \mathbf{\Sigma}||$.

(b) (Davis and Kahan Sinθ Theorem):

$$||\widehat{e}_i - e_i|| \leq \frac{\sqrt{2}||\widehat{\Sigma} - \Sigma||}{min(|\widehat{\lambda}_{i-1} - \lambda_i|, |\lambda_i - \widehat{\lambda}_{i-1}|)}.$$

From this theorem it is immediate that if the sequence $\|\widehat{\Sigma}_n - \Sigma\| \to 0$, then both the *eigenvalues* and *eigenvectors* of $\widehat{\Sigma}_n$ converge to those of Σ. In particular, consistency or convergence of a covariance matrix estimator in the operator norm guarantees the convergence of their PCs loadings (Johnstone and Lu, 2009).

3.3 STRUCTURED COVARIANCE MATRICES

A few examples of commonly used structured covariance matrices involving one or two parameters such as compound symmetry (CS), moving average (MA), and autoregressive (AR) models are introduced in this section. Their positive definiteness is established by computing their eigenvalues in some manners. Such covariance matrices are used often in longitudinal data analysis, engineering, finance and economics, just to name a few.

Example 20 (Compound Symmetry) *The compound symmetry (intra-block, exchangeable, uniform, $\cdots$) covariance matrix*

$$\Sigma = \sigma^2 R = \sigma^2 \begin{pmatrix} 1 & \rho & \cdots & \rho \\ \rho & 1 & \cdots & \rho \\ \vdots & \vdots & \ddots & \vdots \\ \rho & \rho & \cdots & 1 \end{pmatrix} = \sigma^2 \left[(1-\rho)I + \rho \mathbf{1}_n \mathbf{1}_n'\right]$$

has only two parameters σ^2 and ρ. For any vector $\mathbf{c}$, it can be shown that $\mathbf{c}'R\mathbf{c} = (1-\rho)\sum_{i=1}^{p}(c_i - \bar{c})^2 + n\bar{c}^2[1+(p-1)\rho]$, where $\bar{c}$ is the average of the entries of $\mathbf{c}$. Then, a sufficient condition for it to be nonnegative definite is that $1+(p-1)\rho \geq 0$ or $-(p-1)^{-1} \leq \rho \leq 1$. Alternatively, its p eigenvalues can be computed and divided into two groups. For example, when ρ is positive, the largest eigenvalue is

$$\lambda_1 = 1 + (p-1)\rho,$$

with the associated eigenvector (Johnson and Wichern, 2008, Chapter 8),

$$\mathbf{e}_1 = \frac{1}{\sqrt{p}}(1, \cdots, 1)' = \frac{1}{\sqrt{p}}\mathbf{1}_p.$$

The remaining $(p-1)$ eigenvalues are

$$\lambda_2 = \lambda_3 = \cdots = \lambda_p = 1 - \rho,$$

so that the eigenvalues are all nonnegative only if $1 \geq \rho > -(p-1)^{-1}$.

Example 21 (Huynh–Feldt Structure) *This covariance matrix is more general than the compound symmetry matrix and is given by*

$$\boldsymbol{\Sigma} = \sigma^2(\alpha I + a\mathbf{1}_p' + \mathbf{1}_p a'),$$

where $a = (a_1, \ldots, a_p)'$ *and* α *is a scalar to be chosen so that* $\boldsymbol{\Sigma}$ *is positive definite. The* p *eigenvalues of the matrix* $a\mathbf{1}_p' + \mathbf{1}_p a'$ *are* $\lambda_1 = \mathbf{1}_p' a - \sqrt{pa'a}$, $\lambda_2 = \ldots = \lambda_p = 0$, *so the smallest eigenvalue is negative. Thus,* Σ *is nonnegative definite provided that*

$$\alpha > \mathbf{1}_p' a - \sqrt{pa'a}.$$

Example 22 (The One-Dependent Covariance Structure) *A one-dependent covariance matrix has a tri-diagonal structure*

$$\boldsymbol{\Sigma} = \sigma^2 \begin{pmatrix} a & b & 0 & \cdots & 0 \\ b & a & b & \ddots & \vdots \\ \vdots & \ddots & \ddots & \ddots & 0 \\ \vdots & & \ddots & \ddots & b \\ 0 & \cdots & 0 & b & a \end{pmatrix}.$$

Its eigenvalues and eigenvectors are given by

$$\lambda_k = a + 2b\cos k\theta, \; k = 1, \cdots, p$$

and

$$\mathbf{e}_k = (\sin k\theta, \sin 2k\theta, \cdots, \sin pk\theta)',$$

where $\theta = \pi/(p+1)$.

Example 23 (The AR(1) Structure) *The highly popular AR(1) covariance matrix with two parameters* σ^2 *and* ρ *given by*

$$\boldsymbol{\Sigma} = \sigma^2 \begin{pmatrix} 1 & \rho & \cdots & \rho^{p-1} \\ \rho & 1 & \cdots & \rho^{p-1} \\ \vdots & \vdots & \ddots & \vdots \\ \rho^{p-1} & \rho^{p-2} & \cdots & 1 \end{pmatrix}, \quad -1 < \rho < 1, \tag{3.15}$$

is used in many areas of application. It is not easy to show directly its positive definiteness. However, it is much easier to construct a random vector $\mathbf{Y} = (Y_1, \ldots, Y_p)'$ *with* $\mathbf{\Sigma}$ *as its covariance matrix. To this end, let*

$$Y_t = \sum_{k=0}^{\infty} \rho^k \varepsilon_{t-k},$$

where $\{\varepsilon_t\}$ *is a sequence of i.i.d. random variables with* $E(\varepsilon_t) = 0$ *and* $var(\varepsilon_t) = \sigma_\varepsilon^2$. *Then, from routine calculations it follows that*

$$\begin{aligned} cov(Y_i, Y_j) &= E \sum_{k=0}^{\infty} \rho^k \varepsilon_{i-k} \sum_{\ell=0}^{\infty} \rho^\ell \varepsilon_{j-\ell}, \\ &= \sum_{k=0}^{\infty} \sum_{\ell=0}^{\infty} \rho^{k+\ell} E(\varepsilon_{i-k}\varepsilon_{j-\ell}), \\ &= \sigma_\varepsilon^2 \sum_{\ell=0}^{\infty} \rho^{i-j+2\ell}, \\ &= \sigma_\varepsilon^2 (1-\rho^2)^{-1} \rho^{i-j} = \sigma^2 \rho^{i-j}, \end{aligned}$$

where $\sigma^2 = \sigma_\varepsilon^2(1-\rho^2)^{-1}$. *This is precisely the* (i, j)*th entry of* $\mathbf{\Sigma}$.

Furthermore, from the definition of Y_t *it follows that*

$$Y_t = \varepsilon_t + \sum_{k=1}^{\infty} \rho^k \varepsilon_{t-k} = \varepsilon_t + \rho \sum_{k=0}^{\infty} \rho^k \varepsilon_{t-1-k} = \varepsilon_t + \rho Y_{t-1},$$

so that the process $\{Y_t\}$ *is an* autoregressive process *of order one, or AR(1) for short and the matrix in (3.15) is usually referred to as the AR(1) covariance structure.*

It can be shown that the eigenvalues λ *of* Σ *(for* $\sigma^2 = 1$*) are the* p *roots of*

$$\frac{(-\lambda\rho)^p}{1-\rho^2} \left\{ \frac{\sin(p+1)\theta}{\sin\theta} - 2\rho \frac{\sin p\theta}{\sin\theta} + \rho^2 \frac{\sin(p-1)\theta}{\sin\theta} \right\} = 0, \tag{3.16}$$

where λ *and* θ *are related via*

$$-2\rho\lambda\cos\theta = 1 - \lambda - \rho^2(1+\lambda). \tag{3.17}$$

Writing the three sine terms in the above expression in terms of $\alpha = \cos\theta$ *reduces (3.16) to a polynomial of degree* p *in* α *with* p *real and distinct zeros* $\alpha_k = \cos\theta_k$ *where*

$$0 < \theta_1 < \theta_2 < \cdots < \theta_p < \pi.$$

Then, the eigenvalues of Σ *computed from (3.17) are*

$$\lambda_k = \frac{1-\rho^2}{1-2\rho\cos\theta_k+\rho^2}, \quad k = 1, \cdots, p.$$

Explicit evaluation of the θ_k*'s does not seem to be possible (see Grenander and Szegö, 1984, p. 70).*

3.4 FUNCTIONS OF A COVARIANCE MATRIX

The square roots, inverse, and logarithm of a covariance matrix are important in understanding some of the deeper aspects of $\boldsymbol{\Sigma}$. In this section, the roles of the spectral decomposition of a symmetric matrix in defining and computing such functions of a positive definite matrix are presented.

As an example, if P is an orthogonal matrix, it follows from (3.13) that $A^2 = AA = P\Lambda P'P\Lambda P' = P\Lambda^2 P'$, and hence

$$A^m = P\Lambda^m P',$$

for any positive integer m where $\Lambda^m = \text{diag}(\lambda_1^m, \ldots, \lambda_n^m)$. Interestingly, the same argument works for noninteger and negative powers. For a nonnegative definite matrix A and $m = \frac{1}{2}$, setting

$$A^{1/2} = \sqrt{\lambda_1}e_1e_1' + \cdots + \sqrt{\lambda_n}e_ne_n' = P\Lambda^{1/2}P',$$

it follows that

$$A^{1/2}A^{1/2} = P\Lambda^{1/2}P'P\Lambda^{1/2}P' = P\Lambda P' = A,$$

so that $A^{1/2}$ is, indeed, a *square root* of the matrix A. Similarly, the *inverse* of a nonsingular symmetric matrix A is given by

$$A^{-1} = \sum_{i=1}^{p} \lambda_i^{-1} e_i e_i' = P\Lambda^{-1}P'.$$

More generally, for any function $f(\cdot)$ with the Taylor expansion

$$f(x) = \sum_{k=0}^{\infty} f^{(k)}(0)\frac{x^k}{k!},$$

and an appropriate symmetric matrix A with the spectral decomposition (3.13), one defines

$$f(A) = \sum_{k=0}^{\infty} f^{(k)}(0)\frac{A^k}{k!} = P \sum_{k=0}^{\infty} f^{(k)}(0)\frac{\Lambda^k}{k!} P' = Pf(\Lambda)P', \tag{3.18}$$

where $f(\Lambda) = \text{diag}(f(\lambda_1), \ldots, f(\lambda_p))$. Using (3.18), for any $p \times p$ symmetric matrix A its matrix exponential e^A is defined by the power series expansion

$$e^A = I + \frac{A}{1!} + \frac{A^2}{2!} + \cdots,$$

where using the spectral decomposition of A, it reduces to

$$e^A = P\text{diag}(e^{\lambda_1}, \cdots, e^{\lambda_p})P'.$$

Of particular interest in covariance modeling is the log of a covariance matrix. The *logarithm of a matrix* A, in symbols $\log A$, is defined using (3.18) by

$$\log A = P(\log \Lambda)P'. \tag{3.19}$$

For a diagonal matrix Λ with positive diagonal entries, its logarithm is the diagonal matrix:

$$\log \Lambda = \text{diag}(\log \lambda_1, \cdots, \log \lambda_p),$$

so that the relationships among entries of Λ and $\log \Lambda$ are pretty simple. However, this is not the case for the entries of A and $\log A$ since

$$A = e^{\log A} \tag{3.20}$$

(see Example 25). From (3.19) the logarithm of a positive definite matrix is simply a symmetric matrix, so that the log transformation removes the notorious positive-definiteness constraint on a covariance matrix. This is of particular interest in modeling covariance matrices using covariates.

Lemma 4 *A matrix A is positive definite, if and only if* $\log A$ *is a real symmetric matrix.*

Example 24 *For the 2×2 positive definite matrix*

$$A = \begin{pmatrix} 2 & 1 \\ 1 & 2 \end{pmatrix},$$

its spectral decomposition is

$$A = \frac{1}{2}\begin{pmatrix} 1 & 1 \\ 1 & -1 \end{pmatrix}\begin{pmatrix} 3 & 0 \\ 0 & 1 \end{pmatrix}\begin{pmatrix} 1 & 1 \\ 1 & -1 \end{pmatrix}.$$

Thus, its symmetric square root and logarithm are given by

$$A^{1/2} = \frac{1}{2}\begin{pmatrix} 1 & 1 \\ 1 & -1 \end{pmatrix}\begin{pmatrix} \sqrt{3} & 0 \\ 0 & 1 \end{pmatrix}\begin{pmatrix} 1 & 1 \\ 1 & -1 \end{pmatrix} = \frac{1}{2}\begin{pmatrix} \sqrt{3}+1 & \sqrt{3}-1 \\ \sqrt{3}-1 & \sqrt{3}+1 \end{pmatrix};$$

$$\log A = \frac{1}{2}\begin{pmatrix} 1 & 1 \\ 1 & -1 \end{pmatrix}\begin{pmatrix} \log 3 & 0 \\ 0 & 0 \end{pmatrix}\begin{pmatrix} 1 & 1 \\ 1 & -1 \end{pmatrix} = \frac{\log 3}{2}\begin{pmatrix} 1 & 1 \\ 1 & 1 \end{pmatrix}.$$

Note that while A has rank two, its log is singular with rank one.

Example 25 (Computing Entries of $\mathbf{\Sigma}$ from $\log \mathbf{\Sigma}$) *Since the entries of* $\log \mathbf{\Sigma}$ *are unconstrained, they can be modeled using covariates (Chiu et al., 1996). Then, using (3.20) allows one to express entries of* $\mathbf{\Sigma}$ *in terms of those of* $\log \mathbf{\Sigma}$*. The following examples reveal the complexity of the procedure. In fact, since the entries of* $\mathbf{\Sigma}$ *cannot be expressed as simple functions of the new parameters, it is difficult to interpret them statistically.*

(a) **Compound Symmetry**
Set

$$\log \mathbf{\Sigma} = (\alpha - \beta)I_p + \beta J_p,$$

where J_p *is the* $p \times p$ *matrix of 1's. From Example 20, the eigenvalues of* $\log \mathbf{\Sigma}$ *are* $\alpha + (p-1)\beta$ *with the normalized eigenvector* $\mathbf{e}_1 = \frac{1}{\sqrt{p}}1_p$*, and the other eigenvalues of* $\mathbf{\Sigma}$ *are* $\alpha - \beta$ *with multiplicity* $p - 1$*. Thus, with* $\Lambda = diag[\alpha + (p-1)\beta, \alpha - \beta, \ldots, \alpha - \beta]$ *we have*

$$\exp(\Lambda) = e^{\alpha-\beta}I_p + (e^{p\beta} - 1)diag(1, 0, \cdots, 0)$$

and

$$\mathbf{\Sigma} = P\exp(\Lambda)P' = e^{\alpha-\beta}I_p + p^{-1}(e^{p\beta} - 1)J_p$$

has exactly the same form as $\log \mathbf{\Sigma}$ *except that the diagonal entry of the latter* $e^{\alpha-\beta} + e^{p\beta} - 1$ *depends on both* α *and* β*.*

(b) **One-Dependence Structure**
Let

$$\log \mathbf{\Sigma} = \begin{pmatrix} \alpha & \beta & 0 & \cdots & \cdots & 0 \\ \beta & \alpha & \ddots & \ddots & & \vdots \\ 0 & \ddots & \ddots & \ddots & \ddots & \vdots \\ \vdots & \ddots & \ddots & \ddots & \ddots & 0 \\ \vdots & & \ddots & \ddots & \ddots & \beta \\ 0 & \cdots & \cdots & 0 & \beta & \alpha \end{pmatrix},$$

where α and β are arbitrary scalars. The eigenvalues and eigenvectors of $\log \mathbf{\Sigma}$ *are given in Example 22. From* $\mathbf{\Sigma} = P\exp(\Lambda)P'$ *it is evident that its* (i, j)*th entry is of the form* $\sigma_{ij} = \frac{2}{p+1}\sum_{k=1}^{p} \sin ik\theta \sin jk\theta \exp(\alpha + 2\beta\cos k\theta)$, *so that* $\mathbf{\Sigma}$ *no longer has a tri-diagonal or one-dependent structure.*

(c) **The General 2×2 Matrix**
For arbitrary real numbers α, β, and γ, set

$$\log \mathbf{\Sigma} = \begin{pmatrix} \alpha & \beta \\ \beta & \gamma \end{pmatrix}.$$

The eigenvalues and eigenvectors of $\log \mathbf{\Sigma}$ *are*

$$\lambda_1 = \tfrac{1}{2}(\alpha + \gamma - \sqrt{\Delta}), \qquad \mathbf{e}_1 = d_1^{-1}(-\beta, \alpha - \lambda_1)',$$
$$\lambda_2 = \tfrac{1}{2}(\alpha + \gamma + \sqrt{\Delta}), \qquad \mathbf{e}_2 = d_2^{-1}(-\beta, \alpha - \lambda_2)',$$

where $\Delta = (\alpha - \gamma)^2 + 4\beta^2$ *and*

$$d_1^2 = \sqrt{\Delta}\,(\alpha - \lambda_1),\ d_2^2 = \sqrt{\Delta}\,(\lambda_2 - \alpha).$$

Thus, with $\Lambda = diag(\lambda_1, \lambda_2)$ *and* $P = (\mathbf{e}_1, \mathbf{e}_2)$, *the entries of the* 2×2 *covariance matrix*

$$\mathbf{\Sigma} = P\exp(\Lambda)P'$$

are given by

$$\sigma_{11} = \frac{1}{2\sqrt{\Delta}} \exp\left(\frac{\alpha+\gamma}{2}\right)\{\sqrt{\Delta}\, s^+ - (\alpha - \gamma)s^-\},$$
$$\sigma_{22} = \frac{1}{2\sqrt{\Delta}} \exp\left(\frac{\alpha+\gamma}{2}\right)\{\sqrt{\Delta}\, s^+ + (\alpha - \gamma)s^-\},$$

and

$$\sigma_{12} = \sigma_{21} = \frac{\beta}{2\sqrt{\Delta}} \exp\left(\frac{\alpha+\gamma}{2}\right) s^{-},$$

where

$$s^{\pm} = \exp\left(\frac{\sqrt{\Delta}}{2}\right) \pm \exp\left(-\frac{\sqrt{\Delta}}{2}\right).$$

(d) *When in Example 25(c),* $\log \boldsymbol{\Sigma}$ *is modeled using a covariate, say x, as in*

$$\log \boldsymbol{\Sigma} = \begin{pmatrix} \alpha_1 + \alpha_2 x & \alpha_5 \\ \alpha_5 & \alpha_3 + \alpha_4 x \end{pmatrix},$$

then σ_{12}*, the covariance between the two random variables in* $\boldsymbol{Y} = (Y_1, Y_2)'$ *is a rather complicated function of the parameters* α_i*'s:*

$$\sigma_{12} = \frac{\alpha_5}{2\sqrt{\alpha}}\left(e^{\sqrt{\alpha}/2} - e^{-\sqrt{\alpha}/2}\right) \exp\left\{\frac{(\alpha_1+\alpha_3)+(\alpha_2+\alpha_4)^2}{2}\right\},$$

where

$$\alpha = \{(\alpha_1 - \alpha_3) + (\alpha_2 - \alpha_4)x\}^2 + 4\alpha_5^2.$$

It is interesting to note that the sign of σ_{12} *is determined by the sign of* α_5 *which is the (1,2) entry of* $\log \boldsymbol{\Sigma}$.

(e) *Starting from the log-linear model*

$$\log \boldsymbol{\Sigma} = \begin{pmatrix} \alpha_1 & \alpha_3 + \alpha_4 x \\ \alpha_3 + \alpha_4 x & \alpha_2 \end{pmatrix},$$

it follows from Example 25(c) that

$$\sigma_{11} = var(Y_1) = \frac{1}{2\sqrt{\beta}} \exp\left(\frac{\alpha_1+\alpha_2}{2}\right)$$

$$\left[\sqrt{\beta}\left(e^{\frac{\sqrt{\beta}}{2}} + e^{\frac{-\sqrt{\beta}}{2}}\right) + (\alpha_1 - \alpha_2)\left(e^{\frac{\sqrt{\beta}}{2}} - e^{\frac{-\sqrt{\beta}}{2}}\right)\right],$$

where

$$\beta = (\alpha_1 - \alpha_2)^2 + 4(\alpha_3 + \alpha_4 x)^2.$$

Thus, even though the (1,1) entry of $\log \boldsymbol{\Sigma}$ *does not depend on x, the corresponding entry of* $\boldsymbol{\Sigma}$ *does depend on the covariate x.*

3.5 PCA: THE MAXIMUM VARIANCE PROPERTY

The PCA is concerned with explaining the covariance matrix $\mathbf{\Sigma}$ of $\boldsymbol{Y} = (Y_1, \ldots, Y_p)'$ through a few, say k, linear combinations of these variables. The hope is that some new variables $Z_1, \ldots, Z_k$ with maximum variance will help: (i) to provide a simpler and more parsimonious approximation to $\mathbf{\Sigma}$, (ii) to reduce the dimension of the data from p to k where these linear combinations are possibly easier to interpret and visualize when $k = 2, 3$.

The starting point is the p arbitrary linear combinations

$$\begin{aligned}
Z_1 &= \mathbf{c}_1'\boldsymbol{Y} = c_{11}Y_1 + c_{12}Y_2 + \cdots + c_{1p}Y_p,\\
Z_2 &= \mathbf{c}_2'\boldsymbol{Y} = c_{21}Y_1 + c_{22}Y_2 + \cdots + c_{2p}Y_p,\\
&\vdots\\
Z_p &= \mathbf{c}_p'\boldsymbol{Y} = c_{p1}Y_1 + c_{p2}Y_2 + \cdots + c_{pp}Y_p,
\end{aligned}$$

with

$$\begin{aligned}
\operatorname{var}(Z_i) &= \mathbf{c}_i'\mathbf{\Sigma}\mathbf{c}_i,\ i = 1, \cdots, p,\\
\operatorname{cov}(Z_i, Z_j) &= \mathbf{c}_i'\mathbf{\Sigma}\mathbf{c}_j,\ i, j = 1, \cdots, p.
\end{aligned}$$

Then, the *PCs* of $\boldsymbol{Y}$ are those *uncorrelated* random variables $Z_1, \ldots, Z_p$ whose variances are as large as possible. More precisely, the *first principal component* Z_1 is obtained by finding $\mathbf{c}_1$ so that $\operatorname{var}(Z_1) = \mathbf{c}_1'\mathbf{\Sigma}\mathbf{c}_1$ is maximized. Since this variance can be increased by multiplying any $\mathbf{c}_1$ by a constant, we can eliminate this indeterminacy by restricting $\mathbf{c}_1$ to have unit length, that is, $\mathbf{c}_1'\mathbf{c}_1 = 1$. The *second principal component* Z_2 is obtained by finding $\mathbf{c}_2$ so that $\operatorname{var}(Z_2) = \mathbf{c}_2'\mathbf{\Sigma}\mathbf{c}_2$ is maximized subject to

$$\mathbf{c}_2'\mathbf{c}_2 = 1,$$

and

$$\operatorname{cov}(Z_1, Z_2) = \mathbf{c}_1'\mathbf{\Sigma}\mathbf{c}_2 = 0.$$

Finally, the *ith principal component* is obtained by finding $\mathbf{c}_i$ so that $\operatorname{var}(Z_i) = \mathbf{c}_i'\mathbf{\Sigma}\mathbf{c}_i$ is maximized subject to

$$\mathbf{c}_i'\mathbf{c}_i = 1,$$

and

$$\mathbf{c}_j'\mathbf{\Sigma}\mathbf{c}_i = 0, \text{ for } j = 1, 2, \cdots, i-1.$$

In what follows, we show that $\mathbf{c}_i$ is the ith normalized eigenvector of $\mathbf{\Sigma}$ and $\operatorname{var}(\mathbf{c}_i'\boldsymbol{Y}) = \lambda_i$ is its ith largest eigenvalue. Since $\operatorname{var}(\mathbf{c}'\boldsymbol{Y}) = \mathbf{c}'\Sigma\mathbf{c}$ for any vector $\mathbf{c}$, maximizing this variance subject to the constraint $\mathbf{c}'\mathbf{c} = 1$, by using the Lagrange's method of multipliers, amounts to maximizing the function

$$f_1(\mathbf{c}, \lambda) = \mathbf{c}'\Sigma\mathbf{c} - \lambda(\mathbf{c}'\mathbf{c} - 1).$$

Setting to zero the derivative of f_1 with respect to $\mathbf{c}$, one obtains

$$\frac{\partial f_1}{\partial \mathbf{c}} = 2(\boldsymbol{\Sigma}\mathbf{c} - \lambda\mathbf{c}) = 2(\boldsymbol{\Sigma} - \lambda I)\mathbf{c} = 0,$$

and comparing this with (3.10) reveals that $\mathbf{c}$ is an eigenvector corresponding to an eigenvalue λ of $\boldsymbol{\Sigma}$. If $\mathbf{c}_1$ is a normalized eigenvector, then

$$\mathbf{c}_1'\boldsymbol{\Sigma}\mathbf{c}_1 = \lambda\mathbf{c}_1'\mathbf{c}_1 = \lambda,$$

is maximum when λ is taken to be the largest eigenvalue of $\boldsymbol{\Sigma}$. Thus, the *first principal component* Z_1 of $\boldsymbol{Y}$ with the variance $\text{var}(Z_1) = \text{var}(\mathbf{c}_1'\boldsymbol{Y}) = \lambda_1$ corresponds to the first eigenvalue–eigenvector pair $(\lambda_1, \mathbf{e}_1)$ of the covariance matrix $\boldsymbol{\Sigma}$.

The *second principal component* $Z_2 = \mathbf{c}'\boldsymbol{Y}$ is such that $\mathbf{c}'\mathbf{c} = 1$, and Z_1 and Z_2 are uncorrelated, that is,

$$\text{cov}(Z_1, Z_2) = \mathbf{c}_1'\boldsymbol{\Sigma}\mathbf{c} = \mathbf{c}'\boldsymbol{\Sigma}\mathbf{c}_1 = \lambda_1\mathbf{c}'\mathbf{c}_1 = 0,$$

or $\mathbf{c}$ and $\mathbf{c}_1$ must be orthogonal. Finding Z_2 using the Lagrange's multipliers for the two constraints amounts to maximizing the function

$$f_2(\mathbf{c}, \lambda, \gamma) = \mathbf{c}'\boldsymbol{\Sigma}\mathbf{c} - \lambda(\mathbf{c}'\mathbf{c} - 1) - \gamma\mathbf{c}'\boldsymbol{\Sigma}\mathbf{c}_1.$$

Setting to zero its derivative with respect to $\mathbf{c}$, one obtains

$$\frac{\partial f_2}{\partial \mathbf{c}} = 2(\boldsymbol{\Sigma}\mathbf{c} - \lambda\mathbf{c} - \gamma\boldsymbol{\Sigma}\mathbf{c}_1) = 0, \tag{3.21}$$

so that if $\mathbf{c}_2$ is a solution, then premultiplying (3.21) by $\mathbf{c}_1'$, it follows that

$$\mathbf{c}_1'\boldsymbol{\Sigma}\mathbf{c}_2 - \lambda\mathbf{c}_1'\mathbf{c}_2 - \gamma\mathbf{c}_1'\boldsymbol{\Sigma}\mathbf{c}_1 = -\gamma\mathbf{c}_1'\boldsymbol{\Sigma}\mathbf{c}_1 = 0$$

and hence $\gamma = 0$. Thus, from (3.10) $\mathbf{c}_2$ and λ must satisfy

$$\boldsymbol{\Sigma}\mathbf{c}_2 = \lambda\mathbf{c}_2,$$

and $\text{var}(Z_2) = \mathbf{c}_2'\boldsymbol{\Sigma}\mathbf{c}_2 = \lambda$ is the maximum when λ is taken to be λ_2, the second largest eigenvalue of $\boldsymbol{\Sigma}$. This confirms that the second principal component Z_2 corresponds to the second eigenvalue–eigenvector pair $(\lambda_2, \mathbf{e}_2)$. A similar argument establishes the claim made earlier that the *ith principal component* Z_i of $\boldsymbol{Y}$ corresponds to the ith eigenvalue–eigenvector pair $(\lambda_i, \mathbf{e}_i)$ of $\boldsymbol{\Sigma}$.

While the ith eigenvalue λ_i is the variance of the ith principal component Z_i, the statistical meaning of the e_{ij}, the jth entry of the ith normalized eigenvector $\mathbf{e}_i$ is less clear. However, from

$$Z_i = e_{i1}Y_1 + \cdots + e_{ij}Y_j + \cdots + e_{ip}Y_p,$$

it can be seen that e_{ij} measures the importance of the jth variable Y_j to the ith principal component. Consequently, with $\mathbf{c}_j = (0, \ldots, 0, 1, 0, \ldots, 0)$ having a 1 in the jth position, it follows that

$$\text{cov}(\boldsymbol{Y}_j, Z_i) = \text{cov}(\mathbf{c}'\boldsymbol{Y}, \mathbf{e}_i'\boldsymbol{Y}) = \mathbf{c}'\boldsymbol{\Sigma}\mathbf{e}_i = \lambda_i \mathbf{c}'\mathbf{e}_i = \lambda_i e_{ij},$$

so that the correlation coefficient between Z_i and Y_j is given by

$$\rho_{Z_i, Y_j} = \sqrt{\frac{\lambda_i}{\sigma_{jj}}} \cdot e_{ij},$$

which is proportional to e_{ij} and its sign is determined by that of the e_{ij}.

This section has shown the optimality of PCs in terms of it maximum variance property. Generally, a more direct regression-based approach to the PCA is desirable, particularly if the idea of maximizing the variance of uncorrelated linear combinations of the original variables, can be linked to the more familiar regression criterion of minimizing certain error sum of squares (Jong and Kotz, 1999; Shen and Huang, 2008).

3.6 MODIFIED CHOLESKY DECOMPOSITION

In this section, the close connection between the idea of regression and the Cholesky decomposition of a covariance matrix are used to provide unconstrained and statistically meaningful reparameterizations of a positive definite covariance matrix.

The standard Cholesky decomposition of a positive definite matrix encountered in some optimization techniques and matrix computation (Golub and Van Loan, 1996) is of the form

$$\boldsymbol{\Sigma} = CC', \tag{3.22}$$

where $C = (c_{ij})$ is a unique lower triangular matrix with positive diagonal entries.

Example 26 (Computing the Cholesky Factor) *For the matrix $\boldsymbol{\Sigma}$ in Example 17(a), we have*

$$\begin{pmatrix} 2 & 1 \\ 1 & 2 \end{pmatrix} = \begin{pmatrix} c_{11} & 0 \\ c_{21} & c_{22} \end{pmatrix} \begin{pmatrix} c_{11} & c_{21} \\ 0 & c_{22} \end{pmatrix} = \begin{pmatrix} c_{11}^2 & c_{11}c_{21} \\ c_{11}c_{21} & c_{11}^2 + c_{22}^2 \end{pmatrix},$$

which leads to the equations,

$$\begin{cases} c_{11}^2 = 2, \\ c_{11}c_{21} = 1, \\ c_{21}^2 + c_{22}^2 = 2. \end{cases}$$

Assuming that c_{11} and c_{22} are nonnegative, we obtain

$$C = \begin{pmatrix} \sqrt{2} & 0 \\ \frac{1}{\sqrt{2}} & \frac{\sqrt{3}}{2} \end{pmatrix},$$

where the diagonal entries are not unity.

Statistical interpretation of the entries of C is difficult in its present form for a given covariance matrix. However, reducing C to unit lower triangular matrices through multiplication by the inverse of $D_1 = diag(c_{11}, \cdots, c_{pp})$ makes the task of statistical interpretation of the diagonal entries of C and the ensuing unit lower triangular matrix much easier.

Using basic matrix multiplication (3.22) can be rewritten as

$$\boldsymbol{\Sigma} = CD_1^{-1}D_1D_1D_1^{-1}C' = LD_1^2L', \tag{3.23}$$

where $L = CD_1^{-1}$ is obtained from C by dividing the entries of its ith column by c_{ii}. The factorization (3.23), called the *modified Cholesky decomposition* of $\boldsymbol{\Sigma}$, can also be written in the forms

$$T\boldsymbol{\Sigma}T' = D, \boldsymbol{\Sigma}^{-1} = T'D^{-1}T, \tag{3.24}$$

where $D = D_1^2$ and $T = L^{-1}$ is a unit lower triangular matrix. Note that the second identity can be seen as the modified Cholesky decomposition of the precision matrix $\boldsymbol{\Sigma}^{-1}$. Moreover, the first identity in (3.24) is similar to the spectral decomposition, in that $\boldsymbol{\Sigma}$ is diagonalized by a unit lower triangular matrix. However, unlike the constrained entries of the orthogonal matrix of the spectral decomposition, the nonredundant entries of T are unconstrained and statistically meaningful.

The derivation of the decomposition in (3.24) presented next makes it clear that the Cholesky factors T and D are dependent on the *order* of the variables in $\boldsymbol{Y}$. The idea of regression was used in Section 1.6 to show that T and D can be constructed directly by regressing each variable Y_t on its predecessors. In what follows, it is assumed that $\boldsymbol{Y}$ is a random vector with mean zero and a positive definite covariance matrix $\boldsymbol{\Sigma}$. Let $\widehat{Y}_t$ be the linear least-squares predictor of Y_t based on its predecessors $Y_{t-1} \ldots, Y_1$ and $\varepsilon_t = Y_t - \widehat{Y}_t$ be its prediction error with variance $\sigma_t^2 = \text{var}(\varepsilon_t)$. Then, there are unique scalars ϕ_{tj}, so that

$$Y_t = \sum_{j=1}^{t-1} \phi_{tj}Y_j + \varepsilon_t, \; t = 1, \cdots, p. \tag{3.25}$$

The regression coefficients ϕ_{tj}'s in (3.25) will be expressed in terms of the covariance matrix $\boldsymbol{\Sigma}$. For this and other regression-based techniques, we consider more general regression (prediction) problems and partition a mean zero random vector $\boldsymbol{Y}$ into two components $(\boldsymbol{Y}_1', \boldsymbol{Y}_2')'$ of dimensions p_1 and p_2, respectively. Likewise, we partition its covariance and precision matrices conformally as

$$\boldsymbol{\Sigma} = \begin{pmatrix} \boldsymbol{\Sigma}_{11} & \boldsymbol{\Sigma}_{12} \\ \boldsymbol{\Sigma}_{21} & \boldsymbol{\Sigma}_{22} \end{pmatrix}, \quad \boldsymbol{\Sigma}^{-1} = \begin{pmatrix} \boldsymbol{\Sigma}^{11} & \boldsymbol{\Sigma}^{12} \\ \boldsymbol{\Sigma}^{21} & \boldsymbol{\Sigma}^{22} \end{pmatrix}.$$

Some useful relationships among the blocks of $\boldsymbol{\Sigma}$ and $\boldsymbol{\Sigma}^{-1}$ will be obtained by considering the linear least-squares regression (prediction) of $\boldsymbol{Y}_2$ based on $\boldsymbol{Y}_1$. Let the $p_2 \times p_1$ matrix $\Phi_{2|1}$ be the regression coefficients and the vector of residuals be denoted by $\boldsymbol{Y}_{2\cdot1} = \boldsymbol{Y}_2 - \Phi_{2|1}\boldsymbol{Y}_1$. Then, $\Phi_{2|1}$ and the corresponding prediction error covariance matrix are computed using the fact that the vector of residuals $\boldsymbol{Y}_{2\cdot1}$ is uncorrelated with $\boldsymbol{Y}_1$. This leads to

$$\Phi_{2|1} = \boldsymbol{\Sigma}_{21}\boldsymbol{\Sigma}_{11}^{-1}, \tag{3.26}$$

and

$$\text{cov}(\boldsymbol{Y}_{2\cdot1}) = \boldsymbol{\Sigma}_{22} - \boldsymbol{\Sigma}_{21}\boldsymbol{\Sigma}_{11}^{-1}\boldsymbol{\Sigma}_{12} = \boldsymbol{\Sigma}_{22\cdot1}. \tag{3.27}$$

The following lemma provides more details on the derivation of the above formulas and re-expresses the matrix of regression coefficients $\Phi_{2|1}$ and the covariance matrix $\boldsymbol{\Sigma}_{22\cdot1}$ in terms of the corresponding blocks of the precision matrix.

Lemma 5 *With notation as above, we have*

$$\Phi_{2|1} = \boldsymbol{\Sigma}_{21}\boldsymbol{\Sigma}_{11}^{-1} = -(\boldsymbol{\Sigma}^{22})^{-1}\boldsymbol{\Sigma}^{21}, \tag{3.28}$$

and

$$cov(\boldsymbol{Y}_{2\cdot1}) = \boldsymbol{\Sigma}_{22\cdot1} = (\boldsymbol{\Sigma}^{22})^{-1}. \tag{3.29}$$

Proof *Recall that the matrix of regression coefficients $\Phi_{2|1}$ is found so that the vector of residuals $\boldsymbol{Y}_{2\cdot1}$ is uncorrelated with the data $\boldsymbol{Y}_1$:*

$$\begin{aligned} 0 = cov(\boldsymbol{Y}_{2\cdot1}, \boldsymbol{Y}_1) &= cov(\boldsymbol{Y}_2, \boldsymbol{Y}_1) - \Phi_{2|1}cov(\boldsymbol{Y}_1, \boldsymbol{Y}_1) \\ &= \boldsymbol{\Sigma}_{21} - \Phi_{2|1}\boldsymbol{\Sigma}_{11}, \end{aligned} \tag{3.30}$$

or

$$\Phi_{2|1} = \boldsymbol{\Sigma}_{21}\boldsymbol{\Sigma}_{11}^{-1}.$$

The covariance matrix of $\boldsymbol{Y}_{2\cdot 1}$, the regression residuals, denoted by $\boldsymbol{\Sigma}_{22\cdot 1}$, is

$$\begin{aligned}\boldsymbol{\Sigma}_{22\cdot 1} &= \text{cov}(\boldsymbol{Y}_{2\cdot 1}) = \text{cov}(\boldsymbol{Y}_2 - \Phi_{2|1}\boldsymbol{Y}_1,\ \boldsymbol{Y}_2 - \Phi_{2|1}\boldsymbol{Y}_1),\\ &= \text{cov}(\boldsymbol{Y}_2 - \Phi_{2|1}\boldsymbol{Y}_1,\ \boldsymbol{Y}_2),\\ &= \boldsymbol{\Sigma}_{22} - \Phi_{2|1}\boldsymbol{\Sigma}_{12},\\ &= \boldsymbol{\Sigma}_{22} - \boldsymbol{\Sigma}_{21}\boldsymbol{\Sigma}_{11}^{-1}\boldsymbol{\Sigma}_{12}.\end{aligned} \tag{3.31}$$

Since the vector of residuals $\boldsymbol{Y}_{2\cdot 1}$ is a linear transformation of $\boldsymbol{Y}_1$ and $\boldsymbol{Y}_2$, it is evident that

$$\begin{pmatrix}\boldsymbol{Y}_1\\ \boldsymbol{Y}_{2\cdot 1}\end{pmatrix} = T\begin{pmatrix}\boldsymbol{Y}_1\\ \boldsymbol{Y}_2\end{pmatrix}, \tag{3.32}$$

with T a block lower triangular matrix with the identity blocks down the diagonals and $-\Phi_{2|1}$ in the $(2, 1)$ block. The covariance matrix of the left-hand side in (3.32) is by construction the block-diagonal matrix:

$$D = \text{diag}(\boldsymbol{\Sigma}_{11}, \boldsymbol{\Sigma}_{22\cdot 1}). \tag{3.33}$$

Thus, computing the covariance of the linear transformation in (3.32) we obtain

$$T\boldsymbol{\Sigma}T' = D, \tag{3.34}$$

and the precision matrix $\boldsymbol{\Sigma}^{-1}$ has a similar decomposition:

$$\boldsymbol{\Sigma}^{-1} = T'D^{-1}T. \tag{3.35}$$

Important consequences of multiplying the partitioned matrices in the right-hand side of this identity and matching with the $(2, 2)$ and $(1, 2)$ blocks of $\boldsymbol{\Sigma}^{-1}$ are precisely the desired results.

Next, using the above result we express the regression coefficients ϕ_{tj} in (3.25) in terms of the covariance matrix. For a fixed t, $2 \le t \le p$, set $\phi_t = (\phi_{t1}, \cdots, \phi_{t,t-1})'$ and let $\boldsymbol{\Sigma}_t$ be the $(t-1) \times (t-1)$ leading principal minor of $\boldsymbol{\Sigma}$ and $\tilde{\sigma}_t$ be the column vector composed of the first $t-1$ entries of the tth column of $\boldsymbol{\Sigma}$. Then, from (3.26) and (3.27) with $\boldsymbol{Y}_1 = (Y_1, \cdots, Y_{t-1})'$, $\boldsymbol{Y}_2 = Y_t$, it follows that

$$\phi_t = \Sigma_t^{-1}\tilde{\sigma}_t, \sigma_t^2 = \sigma_{tt} - \tilde{\sigma}_t'\Sigma_t^{-1}\tilde{\sigma}_t. \tag{3.36}$$

Since the ϕ_{ij}'s in (3.36) are simply the regression coefficients computed from an unstructured covariance matrix, these coefficients along with $\log\sigma_t^2$ are unconstrained (Pourahmadi, 1999). From (3.25) it is evident that the regression or the orthogonalization process reduces the task of modeling a covariance matrix to that of a sequence of p varying-coefficient and varying-order regression models. Thus, one can bring the familiar regression analysis machinery to handle the unintuitive task of modeling

TABLE 3.1 Regression coefficients and residual variances

Y_1	Y_2	Y_3	$\cdots$	Y_{p-1}	Y_p
1					
ϕ_{21}	1				
ϕ_{31}	ϕ_{32}	1			
$\vdots$	$\vdots$		$\ddots$		
$\vdots$	$\vdots$			$\ddots$	
ϕ_{p1}	ϕ_{p2}	$\cdots$	$\cdots$	$\phi_{p,p-1}$	1
σ_1^2	σ_2^2	$\cdots$	$\cdots$	σ_{p-1}^2	σ_p^2

covariance matrices. Table 3.1 provides a schematic view of the successive variables, their regression coefficients, and residual variances. An important consequence of the decomposition (3.24) is that, for any estimate $(\widehat{T}, \widehat{D})$ of the Cholesky factors, the estimated precision matrix $\widehat{\boldsymbol{\Sigma}}^{-1} = \widehat{T}'\widehat{D}^{-1}\widehat{T}$ is guaranteed to be positive definite.

There is an alternative form of the Cholesky decomposition due to Chen and Dunson (2003) which can also be obtained from (3.22):

$$\boldsymbol{\Sigma} = D_1 \tilde{L}\tilde{L}' D_1,$$

where $\tilde{L} = D_1^{-1}C$ is obtained from C by dividing the entries of its ith row by c_{ii}. This form has proved useful for joint variable selection for fixed and random effects in the linear mixed-efects models, and when the focus is on modeling the correlation matrix, see Pourahmadi (2007a).

There is a long history of the roles of the Cholesky decomposition in statistics. Some early and implicit examples of the use of the Cholesky decomposition in the literature of statistics include Bartlett's (1933) decomposition of a sample covariance matrix; Wright's (1934) path analysis; Roy's (1958) step-down procedures, and Wold's (1960) causal chain models which assume the existence of an a priori order among the p variables of interest. Some of the more explicit uses are in Kalman (1960) for filtering in state–space models and the Gaussian graphical models (Wermuth, 1980). For other uses of Cholesky decomposition in multivariate quality control and related areas see Pourahmadi (2007b).

3.7 LATENT FACTOR MODELS

One of the oldest and most common approaches to dealing with the high-dimensionality problem in covariance modeling is the *statistical factor models* (Anderson, 2003). Historically, factor analysis was developed in psychology in search of unobservable (latent) factors related to intelligence. However, nowadays it is used as a powerful dimension-reduction tool in finance and economics where the common factors can be observable. For example, the popular capital asset pricing model (CAPM) is a single factor model where the proxy for the market is a valued-weighted index like the S&P 500, so that the model has an observable factor (Sharpe, 1970). In multifactor models, the macroeconomic variables used as factors are inflation rate, interest rate, risk premia, and term premia. The factors in the Fama and French (1992) three-factor model are portfolio-based firm characteristics such as the firm size, market value, and the industry type.

The goal of *factor analysis* of a covariance matrix is to identify a few common (latent or observable) factors that can explain the dependence in the data. Algebraically, it amounts to reparameterizing a $p \times p$ covariance matrix $\boldsymbol{\Sigma}$ as in (1.22) with q small so that the factor models lead to a considerable reduction in the number of covariance parameters. Although using factor models does reduce the dimension of the parameter space in many problems, theoretically not every covariance matrix can be decomposed as (1.22) with a small q.

The latent factor model (1.21) is said to be an *orthogonal factor model* if the following assumptions are satisfied:

1. $E(F) = 0$ and $\text{cov}(F) = I_q$;
2. $E(\boldsymbol{\varepsilon}) = 0$ and $\text{cov}(\boldsymbol{\varepsilon}) = \Psi = \text{diag}(\psi_1, \ldots, \psi_p)$;
3. F and $\boldsymbol{\varepsilon}$ are independent so that $\text{cov}(F, \boldsymbol{\varepsilon}) = 0$.

From (1.21) and the above assumptions, it follows that $\boldsymbol{\Sigma}$ admits the following decomposition:

$$\begin{aligned} \boldsymbol{\Sigma} = \text{cov}(\boldsymbol{Y}) &= \text{cov}(LF + \boldsymbol{\varepsilon}), \\ &= LL' + \Psi, \end{aligned} \tag{3.37}$$

where the loading matrix L has the interpretation as the covariance matrix of $\boldsymbol{Y}$ and F:

$$L = \text{cov}(\boldsymbol{Y}, F). \tag{3.38}$$

Therefore, in the orthogonal factor model the entries of $\boldsymbol{\Sigma}$ are related to those of L and Ψ via

$$\begin{aligned} \sigma_{ii} &= \text{var}(Y_i) = \ell_{i1}^2 + \cdots + \ell_{iq}^2 + \psi_i, \\ \sigma_{ij} &= \text{cov}(Y_i, Y_j) = \ell_{i1}\ell_{j1} + \cdots + \ell_{iq}\ell_{jq}. \end{aligned} \tag{3.39}$$

The quantity $\ell_{i1}^2 + \cdots + \ell_{iq}^2 = c_i^2$, the portion of variance of Y_i explained by the q common factors, is called its *communality* and the unexplained portion ψ_i in (3.39) is called its *idiosyncratic*, *uniqueness*, or *specific variance*. Also, it can be seen from (3.38) that ℓ_{ij}, the (i, j)th entry of L, is precisely the covariance between Y_i and the jth factor f_j.

The following example illustrates some of the concepts and quantities introduced thus far.

Example 27 *The* 3×3 *covariance matrix*

$$\Sigma = \begin{pmatrix} 4 & -2 & 3 \\ -2 & 6 & -6 \\ 3 & -6 & 10 \end{pmatrix},$$

admits a decomposition of the form of (3.37):

$$\Sigma = \begin{bmatrix} 1 \\ -2 \\ 3 \end{bmatrix} (1 \ -2 \ 3) + \begin{pmatrix} 3 & 0 & 0 \\ 0 & 2 & 0 \\ 0 & 0 & 1 \end{pmatrix} = LL' + \Psi,$$

which corresponds to an orthogonal factor model with $q = 1$,

$$L = (1, -2, 3)' \text{ and } \Psi = diag(3, 2, 1).$$

Thus, the communality of Y_2 *is*

$$c_2^2 = \ell_{21}^2 = (-2)^2 = 4,$$

and its specific variance is

$$\psi_2 = \sigma_{22} - c_2^2 = 6 - 4 = 2.$$

Similar calculations can be done for the other two variables.

It is instructive to note that the matrix

$$L^* = -L = (-1, 2, -3)'$$

also satisfies the above decomposition which raises the question of identifiability or nonuniqueness of the matrix of factor loadings in factor analysis discussed next.

First, the representation (3.37) is not unique in the sense that for any $p \times p$ orthogonal matrix P, we have

$$Y - \mu = LF + \boldsymbol{\varepsilon} = LPP'F + \boldsymbol{\varepsilon} = L^*F^* + \boldsymbol{\varepsilon}, \tag{3.40}$$

where $L^* = LP$ and $F^* = P'F$ also satisfy the assumptions of an orthogonal factor model. Although this nonuniqueness of the pair (L, F) appears to be a weakness of factor analysis from the mathematical point of view, it can be turned into considerable advantage in some cases so far as interpretation of the latent factors are concerned. In fact, in spite of the nonuniqueness of (L, F) the communalities and specific variances remain unchanged under rotations or orthogonal transformations of the factors. Therefore, it is prudent to search for an orthogonal matrix P so that the common factors have relevant and meaningful substantive interpretation.

A widely used criterion for selecting a suitable rotation is the Kaiser's *varimax* criterion (Johnson and Wichern, 2008, Chapter 9). To introduce it, let $\tilde{\ell}^*_{ij} = \ell^*_{ij}/c_i$ be the rotated coefficients scaled by the (positive) square root of communalities. The varimax procedure selects an orthogonal matrix P that maximizes the quantity

$$V = \frac{1}{p}\sum_{j=1}^{q}\left[\sum_{i=1}^{n}(\tilde{\ell}^*_{ij})^4 - \frac{1}{q}\left(\sum_{i=1}^{q}\tilde{\ell}^{*2}_{ij}\right)^2\right],$$

which amounts to spreading out the squares of the loadings on each factor as much as possible. In other words, the procedure searches for groups of large and negligible coefficients in any column of the rotated matrix of factor loadings.

Unfortunately, not all covariance matrices can be factored as (1.22) for q much smaller than p, as the following example from Johnson and Wichern (2008, Chapter 9) indicates.

Example 28 (Nonexistence of Factor Models) *For the* 3×3 *covariance matrix*

$$\mathbf{\Sigma} = \begin{pmatrix} 1 & 0.9 & 0.7 \\ 0.9 & 1 & 0.4 \\ 0.7 & 0.4 & 1 \end{pmatrix},$$

we use (3.37) to compute the nonredundant entries of L and Ψ for an orthogonal factor model (1.21) with $q = 1$. Matching like entries of the two sides of (3.37), we obtain

$$\begin{aligned} 1 &= \ell_{11}^2 + \psi_1 & 0.90 &= \ell_{11}\ell_{21} & 0.70 &= \ell_{11}\ell_{31} \\ & & 1 &= \ell_{21}^2 + \psi_2 & 0.40 &= \ell_{21}\ell_{31} \\ & & & & 1 &= \ell_{31}^2 + \psi_3 \end{aligned}$$

The (1,3) and (2,3) pair of equations lead to

$$\ell_{21} = \frac{0.40}{0.70}\ell_{11}.$$

Substituting this in the (1,2) equation yields

$$\ell_{11}^2 = 1.575 \text{ or } \ell_{11} = \pm 1.255.$$

But, from $var(f_1) = var(Y_1) = 1$ *and*

$$\ell_{11} = cov(Y_1, f_1) = corr(Y_1, f_1),$$

it follows that the correlation coefficient is greater than 1, indicating that $|\ell_{11}| = 1.255$ *is not a feasible value. Furthermore, the (1,1) equation yields*

$$\psi_1 = 1 - \ell_{11}^2 = 1 - 1.575 = -.575,$$

which is certainly not feasible for $var(\varepsilon_1) = \psi_1$.

This simple example shows that it is possible to get numerical solution to the (3.37), which is outside the feasible parameter range. Such case of improper solution for the pair (L, Ψ) is referred to as the *Heywood case* in the literature of factor analysis. It reveals the difficulty in reparameterizing a covariance matrix $\mathbf{\Sigma}$ in terms of the pair (L, Ψ) for q sufficiently small relative to p.

Estimation of $\theta = (L, \Psi)$: The parameters of the orthogonal factor model can be estimated using either the PCA method or the maximum likelihood method. While the former does not require the normality assumption of the data nor the knowledge of q (the number of common factors), they are required for the latter method.

The PCA Method: This is rather simple to understand and implement. It starts with

$$(\widehat{\lambda}_1, \widehat{e}_1), (\widehat{\lambda}_2, \widehat{e}_2), \cdots, (\widehat{\lambda}_p, \widehat{e}_p),$$

the ordered eigenvalue–eigenvector pairs of the sample covariance matrix. For any $q < p$, the matrix of factor loadings is estimated by

$$\widehat{L} = (\widehat{\ell}_{ij}) = \left[\sqrt{\widehat{\lambda}_1}\widehat{e}_1 \vdots \sqrt{\widehat{\lambda}_2}\widehat{e}_2 \vdots \cdots \vdots \sqrt{\widehat{\lambda}_q}\widehat{e}_q\right], \tag{3.41}$$

and the estimated specific variances are the diagonal elements of the matrix $\mathbf{S} - \widehat{L}\widehat{L}'$, that is,

$$(\widehat{\psi}_1, \cdots, \widehat{\psi}_p) = \text{diag}(\mathbf{S} - \widehat{L}\widehat{L}').$$

Let the error matrix due to approximating $\mathbf{S}$ by $\widehat{L}\widehat{L}' + \widehat{\Psi}$ be defined as

$$E = \mathbf{S} - (\widehat{L}\widehat{L}' + \widehat{\Psi}).$$

How big is this error matrix? It can be shown that

$$||E||_2^2 = \text{tr}EE' \le \widehat{\lambda}_{q+1}^2 + \cdots + \widehat{\lambda}_p^2,$$

that is, the norm of the error is bounded by the sum of squares of the neglected eigenvalues beyond q. Also, it is evident from (3.41) that the earlier estimated factor loadings do not change when one increases q to $q + 1$.

The Maximum Likelihood Method: This method requires that the observations constitute a random sample from a multivariate normal distribution with a nonsingular covariance matrix structured as in (3.37). The log-likelihood function of $\theta = (L, \Psi)$ is of the form

$$L(\theta|\boldsymbol{Y}_1, \cdots, \boldsymbol{Y}_n) = -\frac{n-1}{2}[\log|\Lambda\Lambda' + \Psi| + \mathrm{tr}\{(LL' + \Psi)^{-1}\mathbf{S}\}], \quad (3.42)$$

where $\mathbf{S}$ is the sample covariance matrix. The matrices $\widehat{L}$ and $\widehat{\Psi}$ resulting from maximizing $L(\cdot|\text{data})$ with respect to the entries of L and Ψ are called the maximum likelihood estimates (MLE) of $\theta = (L, \Psi)$.

The indeterminacy of $\widehat{L}$ up to rotation is resolved by constraining $\widehat{L}$ so that

$$\widehat{L}'\widehat{\Psi}^{-1}\widehat{L} = \widehat{\Delta} \quad (3.43)$$

is diagonal. Using this constraint, the likelihood equations can be solved iteratively via the equations

$$\widehat{\Delta}\widehat{L}' = \widehat{L}'\widehat{\Psi}^{-1}(S - \widehat{\Psi}), \ \widehat{\Psi} = \mathrm{diag}(\mathbf{S} - \widehat{L}\widehat{L}'). \quad (3.44)$$

Unfortunately, this iterative method may not converge. To circumvent this difficulty, one may employ a two-step numerical approach to obtaining the MLE of θ. First, for a given Ψ_0 find $\widehat{L} = \widehat{L}_{\Psi_0}$ that maximizes $L(L, \Psi_0|\text{data})$. Then, determine $\widehat{\Psi}$ as that value of Ψ which maximizes $L(\widehat{L}, \Psi|\text{data})$, if necessary one could iterate these two steps. This procedure works well in practice, primarily because the determination of L_{Ψ_0} amounts to computing the p largest eigenvalues and the associated eigenvectors of the matrix $\Psi_0^{-\frac{1}{2}}\mathbf{S}\Psi_0^{-\frac{1}{2}}$.

Factor Scores: In factor analysis, the estimated values of the common factors, called the *factor scores*, are used for diagnostic purposes and inputs to a subsequent analysis. The interest in estimating the common factors is inspired by their conceptual similarity to the problem of estimating the random effects in the mixed-effect models (Searle et al., 1992). In fact, factor scores are not estimates of unknown parameters as in the standard theory of statistics. Instead, they are estimates of values of the unobserved random factors $f_1, \ldots, f_q$. The estimation situation is further complicated by the fact that the parameters $\theta = (L, \Psi)$ are also unknown, and they could easily outnumber the observed data. To circumvent these unusual difficulties, some creative/heuristic methods for estimating the common factors have been suggested. These methods invariably treat the estimated values of $\widehat{\theta} = (\widehat{L}, \widehat{\Psi})$ as if they were the true values, in which case the orthogonal factor model in (1.21) appears like a linear regression model with F as the regression parameter. In this framework, one may use the weighted least-squares (WLS) method to estimate the common factors. That is, choose F to minimize

$$Q(F) = \boldsymbol{\varepsilon}'\widehat{\Psi}^{-1}\boldsymbol{\varepsilon} = (Y - \mu - \widehat{L}F)'\widehat{\Psi}^{-1}(Y - \mu - \widehat{L}F). \quad (3.45)$$

The solution is

$$\widehat{F} = (\widehat{L}'\widehat{\Psi}^{-1}\widehat{L})^{-1}\widehat{L}'\widehat{\Psi}^{-1}(Y - \mu). \tag{3.46}$$

For n subjects, taking $\widehat{\mu} = \bar{Y}$ the factor scores for the jth case is

$$\widehat{F}_j = (\widehat{L}'\widehat{\Psi}^{-1}\widehat{L})^{-1}\widehat{\Psi}^{-1}(Y_j - \bar{Y}),\ j = 1, \cdots, n. \tag{3.47}$$

Recall that if $\widehat{\theta} = (\widehat{L}, \widehat{\Psi})$ are determined by the maximum likelihood method, then these estimates satisfy the uniqueness condition $\widehat{L}'\widehat{\Psi}^{-1}\widehat{L} = \widehat{\Delta}$, a diagonal matrix, and (3.47) reduces to

$$\widehat{F}_j = \widehat{\Delta}^{-1}\widehat{\Psi}^{-1}(Y_j - \bar{Y}),\ j = 1, 2, \cdots, n. \tag{3.48}$$

Orthogonal Factor Models and Partial Correlations: The close connection between orthogonal factor models and the concept of partial correlations can be seen by paraphrasing the purpose of model in (1.21) as the following question: Given p variables Y, do there exist $q(< p)$ variables F such that the partial correlations between every pair of original variables upon elimination of the linear effects of F variables are all zero? We show that a positive answer to this question is equivalent to the orthogonal factor model in (1.21).

First, from the orthogonal factor model in (1.21) it follows that the conditional covariance of Y given F is diagonal, that is

$$\boldsymbol{\Sigma}_{Y|F} = \mathrm{cov}(Y - \mu - LF) = \mathrm{cov}(\varepsilon) = \Psi,$$

and hence the partial correlations between pairs of variables in Y given F are zero. Conversely, suppose there exists F such that the partial correlations between every pair of entries of Y given F is zero. Then, the covariance matrix of the vector of residuals from the linear regression of Y on F is a diagonal matrix. This implies (1.21) with

$$\Lambda = \boldsymbol{\Sigma}_{YF}\Sigma_{FF}^{-1} \quad \text{and} \quad \mathrm{cov}(\varepsilon) = \text{diagonal}.$$

The idea of factor analysis has been extended to nonlinear and nonnormal situations in the past 30 years or so. Also, by reformulating the problem as errors-in-variables regression models the need for rotating the factors and factor loadings has been obviated.

3.8 GLM FOR COVARIANCE MATRICES

A natural way to reduce the large number of covariance parameters is to use covariates as in modeling the mean vector. Progress in modeling and estimation of covariance matrices using covariates has followed a path of developments very much similar to that in regression. It has gone through the familiar phases of linear, log-linear, and generalized linear models (GLMs). The three-stage iterative modeling process based on (i) model formulation, (ii) model estimation, and (iii) model diagnostic has received considerable attention in recent years. The success of generalized linear models hinges on the ability to find unconstrained and statistically meaningful reparameterizations for covariance matrices using spectral and Cholesky decompositions or other means.

In the next three sections, we review the (dis)advantages of linear, log-linear, and GLMs for covariance matrices. The procedures for fitting some of these models are illustrated using an incomplete dataset.

Linear Covariance Models (LCM): The origin of linear models for covariance matrices can be traced to the work of Yule (1927) and Gabriel (1962) and the implicit parameterization of a multivariate normal distribution in terms of entries of either Σ or its inverse. However, Dempster (1972) was the first to recognize the entries of $\boldsymbol{\Sigma}^{-1} = (\sigma^{ij})$ as the canonical parameters of the exponential family of normal distributions. He proposed to select a sparse covariance matrix by identifying zeros in its inverse, and referred to these as *covariance selection* models, it fits the framework of LCM defined next.

The class of LCM is defined via

$$\boldsymbol{\Sigma}^{\pm 1} = \alpha_1 U_1 + \cdots + \alpha_q U_q, \tag{3.49}$$

where U_i's are some known symmetric basis matrices (covariates) and α_i's are unknown parameters (Anderson, 1973). The parameters must be restricted so that the matrix is positive definite. Therefore, it is usually assumed that there is at least a set of coefficients for which $\boldsymbol{\Sigma}^{\pm 1}$ is positive definite.

The model in (3.49) is general enough to include all $p \times p$ covariance matrices. Indeed, for $q = p^2$, any covariance matrix admits the representation:

$$\boldsymbol{\Sigma} = (\sigma_{ij}) = \sum_{i=1}^{p}\sum_{j=1}^{p} \sigma_{ij} U_{ij}, \tag{3.50}$$

where U_{ij} is a $p \times p$ matrix with 1 on the (i, j)th position and 0 elsewhere.

Replacing $\boldsymbol{\Sigma}$ by $\mathbf{S}$ in the left-hand side of (3.49), it can be viewed as a collection of $p(p+1)/2$ linear regression models. The same regression model viewpoint holds with the precision matrix on the left-hand side. A major drawback of (3.49) and (3.50) is the constraint on the coefficients which makes the estimation rather difficult.

For $\boldsymbol{Y}_1, \cdots, \boldsymbol{Y}_n$ a sample of size n from a normal distribution with covariance matrix $\boldsymbol{\Sigma}$ modeled as in (3.49), the score equations for the MLE of the α_j's are

$$\mathrm{tr}\boldsymbol{\Sigma}^{-1}U_i - \mathrm{tr}\mathbf{S}\boldsymbol{\Sigma}^{-1}U_i\boldsymbol{\Sigma}^{-1} = 0, i = 1, \cdots, q. \tag{3.51}$$

These can be solved by an iterative method where in each step a set of linear equations is solved. Anderson (1973, 2003) showed that if consistent estimates of $\alpha_1, \cdots, \alpha_q$ are used as initial values for the coefficients of the linear equations, then its solution or the MLE is asymptotically efficient for large n. Szatrowski (1980) gives necessary and sufficient conditions for the existence of explicit MLE, and the convergence of the iterative procedure proposed by Anderson (1973) in one iteration. In fact, Szatrowski (1980) showed that using a linear covariance model for $\boldsymbol{\Sigma}$, the MLE of the coefficient vector has an explicit representation, that is, a vector of known linear combinations of elements of the sample covariance matrix $\mathbf{S}$, if and only if $\boldsymbol{\Sigma}^{-1}$ has the same LCM pattern. This happens, for example, when $\boldsymbol{\Sigma}$ has a compound symmetry (exchangeable) structure.

An excellent review of the MLE procedures for the model in (3.49) and their applications to the problem of testing homogeneity of the covariance matrices of several dependent multivariate normals are presented in Jiang et al. (1999). They derive a likelihood ratio test, and show how to compute the MLE of $\boldsymbol{\Sigma}$, in both the restricted (null) and unrestricted (alternative) parameter spaces using SAS PROC MIXED software with several examples.

When the LCM (3.49) is viewed as a linear function of its parameters α_i's, then it is called a *linear pencil* (Golub and Van Loan, 1996, p. 394). Note that the set of α_i's for which the linear pencil is nonnegative definite matrix is a convex subset of R^q. In semidefinite programming, one minimizes a linear function subject to the constraint that an affine combination of symmetric matrices is positive semidefinite (Boyd and Vandenberghe, 2011). It unifies several standard problems (e.g., linear and quadratic programming) and has many applications in engineering and combinatorial optimization. When U_i's are diagonal matrices, then one is dealing with a linear programming problem. Adding the requirement of $\boldsymbol{\Sigma}$ being low-rank, then the problem is called *compressed sensing*, see (2.15).

The notion of covariance regression introduced by Hoff and Niu (2011) is in the spirit of (3.49), but unlike the LCM the covariance matrix is quadratic in the covariates, and positive definiteness is guaranteed through the special construction. More precisely, a covariance regression model is of the form

$$\boldsymbol{\Sigma}(x) = A + Bxx'B', x \in R^p,$$

where A is a positive definite and B is a real matrix. This model has interpretable parameters and is related to latent factor models. However, it has some key limitations like scaling to large dimensions and flexibility based on the parametric approach. Specifically, the model restricts the difference between $\boldsymbol{\Sigma}(x)$ and the baseline matrix A to be rank 1. Higher-rank models can be considered via adding additional quadratic terms, but this dramatically increases the number of parameters. A more general class of covariance regression models introduced by Fox and Dunson (2011) relies on the latent factor model and allows the matrix of factor loadings L in (3.37) to depend on the covariates.

Log-Linear Covariance Models: In analogy with the use of log-linear models to handle variance heterogeneity in regression analysis where the variability depends on some predictors, a plausible way to remove the constraint on α_i's in (3.49) is to work with the logarithm of a covariance matrix. The key fact needed here is that for a general covariance matrix with the spectral decomposition $\boldsymbol{\Sigma} = P\Lambda P'$, its *matricial logarithm* defined by $\log \boldsymbol{\Sigma} = P \log \Lambda P'$ is simply a symmetric matrix with unconstrained entries taking values in $(-\infty, \infty)$ (see Section 3.4).

This idea has been pursued by Leonard and Hsu (1992) and Chiu et al. (1996) who introduced the *log-linear covariance* models:

$$\log \boldsymbol{\Sigma} = \alpha_1 U_1 + \cdots + \alpha_q U_q, \tag{3.52}$$

where U_i's are known matrices as before and the α_i's are now unconstrained. However, since $\log \boldsymbol{\Sigma}$ is a highly nonlinear operation on $\boldsymbol{\Sigma}$ (see the examples in Section 3.4), the α_i's lack statistical interpretation (Brown et al., 1994). Fortunately, for $\boldsymbol{\Sigma}$ diagonal since $\log \boldsymbol{\Sigma} = \text{diag}(\log \sigma_{11}, \ldots, \log \sigma_{pp})$ (3.52) amounts to log-linear models for heterogeneous variances which has a long history in econometrics and other areas (see Carroll and Ruppert, 1988, and references therein).

Maximum likelihood estimation procedures to estimate the parameters in (3.52) and their asymptotic properties are studied in Chiu et al. (1996) along with the analysis of two real datasets. Given the flexibility of the log-linear models, one would expect them to be used widely in practice. An interesting application to spatial autoregressive (SAR) models and some of its computational advantages are discussed in LeSage and Pace (2007).

3.9 GLM VIA THE CHOLESKY DECOMPOSITION

In this section, the constraint and lack of interpretation of α_i's in (3.49) and (3.52) are removed simultaneously using the Cholesky decomposition of a covariance matrix introduced in Section 3.6. This opens up the possibility of developing a bona fide GLM methodology for the covariance (precision) matrix in terms of covariates. Model formulation and maximum likelihood estimation (MLE) of the parameters are presented in this and the next sections. An important consequence of the approach based on the modified Cholesky decomposition is that for any estimate of the Cholesky factors, the estimated precision matrix $\widehat{\boldsymbol{\Sigma}}^{-1} = \widehat{T}'\widehat{D}^{-1}\widehat{T}$ is guaranteed to be positive definite, see (3.24).

For model formulation, recall that for an unstructured covariance matrix $\boldsymbol{\Sigma}$, the nonredundant entries of its components $(T, \log D)$ in (3.24) obtained from $\boldsymbol{\Sigma}^{-1}$ are unconstrained. Thus, following the GLM tradition it is plausible to write parameteric models for them using covariates (Pourahmadi, 1999). We introduce graphical tools which may suggest parametric models for ϕ_{tj} and $\log \sigma_t^2$, for $t = 1, \ldots, p$; $j = 1, \ldots, t-1$:

$$\log \sigma_t^2 = z_t'\lambda, \quad \phi_{tj} = z_{tj}'\gamma, \tag{3.53}$$

where z_t, z_{tj} are $q \times 1$ and $d \times 1$ vectors of known covariates, respectively. The vectors $\lambda = (\lambda_1, \ldots, \lambda_q)'$ and $\gamma = (\gamma_1, \ldots, \gamma_d)'$ are the parameters related to the innovation variances and dependence in $\boldsymbol{Y}$, respectively.

Starting from the sample covariance matrix $\mathbf{S}$ one can construct a host of matrices capable of revealing aspects of the dependence structure in the data: notable examples include matrices of sample correlations, partial correlations, adjusted partial correlations (Zimmerman, 2000), and the modified Cholesky decomposition. Graphical methods based on some of these matrices are useful in revealing the importance of certain functions of time and lag as covariates. To identify some of the covariates for pairwise correlation structures, it is prudent to standardize the responses by subtracting and dividing by their sample means and standard deviations, respectively. Then, one may construct the *ordinary scatterplot matrix* (OSM): a two-dimensional array of pairwise scatterplots of the standardized responses. As a graphical counterpart of the sample correlation matrix, the OSM does provide valuable insight into the pairwise dependence structure of the data and other features like outliers, nonlinearities, and clustering of observations. Still, the OSM and sample correlation matrix are limited in revealing the extent of dependence of more than two variables at a time.

Fortunately, there are two complementary tools designed to assess the strength of dependence of several variables at a time. The first is the *partial regression on intervenors scatterplot matrix* (PRISM), which consists of pairwise partial scatterplots of the standardized responses (Zimmerman, 2000). Like the OSM, it is the graphical counterpart of the partial correlation coefficients between two variables adjusted for the linear effect of the intervening variables. For more information on using the PRISM in informal model formulation for longitudinal data, see Zimmerman and Núñez Antón (2009, Chapter 4). The second is the *regressograms* (Pourahmadi, 1999), which are suitable plots of the regression coefficients and the prediction error variances when longitudinal measurements are regressed on their predecessors. In fact, the forms of any postulated model should capture the patterns of two suitable plots corresponding to the two components of the modified Cholesky decomposition of Σ. These plots are referred to collectively as the regressograms (plural) of Σ (Pourahmadi, 1999). The first graph called the *regressogram* (singular) is the plot of the same-lag regression coefficients against their lags, that is, the plot of $\{\phi_{t,t-j}\}$ or the jth subdiagonal entries of T versus the lags $j = 1, \cdots, n-1$. The second graph called the *innovariogram* (Zimmerman and Núñez Antón, 2009, p. 106) plots log σ_t^2 versus the times $t = 1, \ldots, n$. These plots may have recognizable patterns which lead to parsimonious models for (T, D) like low-degree polynomials in the lag and time.

Like scatterplots in regression analysis, correlograms in time series analysis, and variograms in spatial data analysis, regressograms allow formulating parsimonious parametric models for the dependence structure in longitudinal data. It is designed to avoid selecting a covariance matrix from a long menu without carefully examining the data.

The regressograms may suggest parametric models for ϕ_{tj} and $\log \sigma_t^2$ as in (3.53). The most common covariates used in the analysis of several real longitudinal (Pourahmadi, 1999; Pan and MacKenzie, 2003; Ye and Pan, 2006; Leng et al., 2010) are in terms of powers of times and lags:

$$z_t = (1, t, t^2, \cdots, t^{q-1})',$$
$$z_{tj} = (1, t-j, (t-j)^2, \cdots, (t-j)^{d-1})'.$$

A remarkable feature of (3.53) is its flexibility in reducing the potentially high-dimensional and constrained parameters of Σ or the precision matrix to $q + d$ unconstrained parameters λ and γ. Furthermore, one can rely on standard variable selection tools like AIC to identify models such as (3.53) for the data; for more details on the use of AIC and BIC in this context, see Pan and MacKenzie (2003). Interestingly, Ye and Pan (2006) employ such parametrized models for covariance matrices in the context of the popular generalized estimating equations (GEE) for longitudinal data (Liang and Zeger, 1986).

For model estimation or computing the MLE of the parameters, one starts with minus twice the log-likelihood function. For a sample $\boldsymbol{Y}_1, \cdots, \boldsymbol{Y}_n$ from a normal population with the mean vector μ and the common covariance $\boldsymbol{\Sigma}$, except for a constant, is given by

$$
\begin{aligned}
-2l(\boldsymbol{\Sigma}) &= \log|\boldsymbol{\Sigma}| + \sum_{i=1}^{n}(\boldsymbol{Y}_i - \boldsymbol{\mu})'\boldsymbol{\Sigma}^{-1}(\boldsymbol{Y}_i - \boldsymbol{\mu}), \\
&= n\log|D| + n\mathrm{tr}\boldsymbol{\Sigma}^{-1}\mathbf{S}, \\
&= n\log|D| + n\mathrm{tr}D^{-1}T\mathbf{S}T',
\end{aligned}
$$

where $\mathbf{S} = \frac{1}{n}\sum_{i=1}^{n}(\boldsymbol{Y}_i - \bar{\boldsymbol{Y}})(\boldsymbol{Y}_i - \bar{\boldsymbol{Y}})'$ is the sample covariance matrix. The last two equalities are obtained by replacing $\boldsymbol{\Sigma}^{-1}$ from (3.22)–(3.24) and basic matrix operations involving trace of a matrix. Note that the above likelihood is quadratic in T, thus for a given D the MLE of ϕ_{tj}'s has a closed form, the same is true for the MLE of D for a given T (Pourahmadi, 2000; Huang et al., 2007).

An algorithm for computing the MLE of the parameters (γ, λ) using the iterative Newton–Raphson algorithm with Fisher scoring is given in Pourahmadi (2000) along with the asymptotic properties of the estimators. An unexpected result is the asymptotic orthogonality of the MLE of the parameters λ and γ, in the sense that their Fisher information matrix is block-diagonal (see Ye and Pan, 2006; Pourahmadi, 2004, and references therein). When the assumption of normality is questionable like when the data exhibit thick tails, then a multivariate t-distribution might be a reasonable alternative.

The unconstrained nature of the lower triangular matrix T of the Cholesky decomposition of a covariance matrix $\boldsymbol{\Sigma}$ makes it ideal for nonparametric and semiparametric estimation. Wu and Pourahmadi (2003) have used local polynomial estimators to smooth the subdiagonals of T, the smoothing along the subdiagonals was motivated by the similarity of the regressions in (3.25) to the varying-coefficients autoregressions in time series analysis:

$$
\sum_{j=0}^{m} f_{j,p}(t/p)Y_{t-j} = \sigma_p(t/p)\varepsilon_t, t = 0, 1, 2, \cdots,
$$

where $f_{0,p}(\cdot) = 1$, $f_{j,p}(\cdot)$, $1 \le j \le m$, and $\sigma_p(\cdot)$ are continuous functions on [0, 1] and $\{\varepsilon_t\}$ is a sequence of independent random variables each with mean 0 and variance 1. This analogy and the fact that the matrix T for stationary autoregressions have nearly constant entries along subdiagonals suggest taking the subdiagonals of T to be realizations of some smooth univariate functions:

$$\phi_{t,t-j} = f_{j,p}(t/p)\ , \ \sigma_t = \sigma_p(t/p).$$

The details of smoothing and selection of the order m of the autoregression using AIC and a simulation study comparing the performance of the sample covariance matrix to the smoothed estimators are given in Wu and Pourahmadi (2003). Huang et al. (2007) have proposed a more direct and efficient approach using splines to smooth the subdiagonals of T, whereas Leng et al. (2010) estimate a covariance matrix by writing linear models for T and semiparametric models for D. The use of the Cholesky decomposition in estimating the spectral density matrix of a multivariate time series is discussed in Dai and Guo (2004).

3.10 GLM FOR INCOMPLETE LONGITUDINAL DATA

The GLM setup of the previous section encounters the problem of incoherency of the (auto)regression coefficients and innovation variances when the longitudinal data are unbalanced and covariates are used. The problem is addressed in this section. First, we set up the notation for the incomplete data framework. It is convenient to assume that a fixed number of measurements are to be collected at a common set of times for all subjects with a common *grand covariance matrix* $\mathbf{\Sigma}$. However, for some reasons not all responses are observed for all subjects so that a generic subject i's measurements will have a covariance matrix $\mathbf{\Sigma}_i$ which is a principal minor of $\mathbf{\Sigma}$.

3.10.1 The Incoherency Problem in Incomplete Longitudinal Data

Let the vector of repeated measures $\boldsymbol{Y}_i$ of subject i collected at completely irregular times $t_{ij},\ j = 1, \cdots, p_i$, follow a zero mean multivariate normal distribution with covariance matrix $\mathbf{\Sigma}_i$. The modified Cholesky decomposition gives $T_i \Sigma_i T_i' = D_i$, where T_i is a unit lower triangular matrix whose below-diagonal entries are the negatives of the autoregressive coefficients, ϕ_{itj}, in $\widehat{Y}_{it} = \sum_{j=1}^{t-1} \phi_{itj} Y_{ij}$, and D_i is a diagonal matrix whose diagonal entries σ_{it}^2's are the innovation variances of the autoregressions. A GLM for $\mathbf{\Sigma}_i$ can be built for each subject by relating the autoregressive parameters ϕ_{itj} and the log innovation variances $\log \sigma_{it}^2$ to some covariates as

$$\phi_{itj} = z_{itj}' \gamma_i \quad \text{and} \quad \log \sigma_{it}^2 = z_{it}' \lambda_i, \quad 1 \le j \le t-1, 1 \le t \le p, \tag{3.54}$$

where z_{itj} and z_{it} are as before the covariates for the covariance matrices, and $\gamma_i \in R^{q_i}$ and $\lambda_i \in R^{d_i}$ are the corresponding regression parameters which have different dimensions for different subjects. The covariates in (3.54) are usually of the form

$$\begin{aligned} z_{itj} = & \quad (1, (t_{it} - t_{ij}), (t_{it} - t_{ij})^2, \ldots, (t_{it} - t_{ij})^{q-1})', \\ z_{it} = & \quad (1, t_{it}, t_{it}^2, \ldots, t_{it}^{d-1})'. \end{aligned} \tag{3.55}$$

Working with this general class of models gives rise to the following two statistical issues:

- Estimation of γ_i and λ_i based on a single vector $\boldsymbol{Y}_i$ is impossible unless p_i is large or a sort of stationarity assumption is imposed. In other words, there are too many parameters to be estimated using a few observations and one cannot borrow strength from the other subjects.
- Even if these parameters are assumed to be the same for all subjects so that one may borrow strength from other subjects, there remains a problem of interpretation or incoherency of the parameters explained next.

The following example shows the nature of the incoherency problem when the data are unbalanced and one employs a naive method to handle a gap in longitudinal data.

Example 29 *Consider the simple model,* $Y_{it} = \phi Y_{it-1} + \epsilon_{it}$, *for* $t = 2, 3, 4$ *with* $Y_{i1} = \epsilon_{i1}$ *and* $\epsilon_i \sim N_4(0, I)$ *so that for a completely observed subject we have* $D = I_4$ *with the following structures for* T *and* Σ*:*

$$T = \begin{pmatrix} 1 & 0 & 0 & 0 \\ -\phi & 1 & 0 & 0 \\ 0 & -\phi & 1 & 0 \\ 0 & 0 & -\phi & 1 \end{pmatrix},$$

$$\Sigma = \begin{pmatrix} 1 & \phi & \phi^2 & \phi^3 \\ \phi & 1+\phi^2 & \phi^2+\phi^3 & \phi^3+\phi^4 \\ \phi^2 & \phi^2+\phi^3 & 1+\phi^2+\phi^4 & \phi+\phi^3+\phi^5 \\ \phi^3 & \phi^3+\phi^4 & \phi+\phi^3+\phi^5 & 1+\phi^2+\phi^4+\phi^6 \end{pmatrix}.$$

Now, consider two subjects where Subject 1 has three measurements made at times 1, 2, 4 and Subject 2 has measurements at times 1, 3, 4. We obtain $\boldsymbol{\Sigma}_1$ *by deleting the third row and column of* $\boldsymbol{\Sigma}$*, similarly* $\boldsymbol{\Sigma}_2$ *is obtained by deletion of the second row and column of the* $\boldsymbol{\Sigma}$*. The factors of the modified Cholesky decompositions of these matrices are*

$$T_1 = \begin{pmatrix} 1 & 0 & 0 \\ -\phi & 1 & 0 \\ 0 & -\phi^2 & 1 \end{pmatrix}, \quad D_1 = \begin{pmatrix} 1 & 0 & 0 \\ 0 & 1 & 0 \\ 0 & 0 & 1+\phi^2 \end{pmatrix},$$

$$T_2 = \begin{pmatrix} 1 & 0 & 0 \\ -\phi^2 & 1 & 0 \\ 0 & -\phi & 1 \end{pmatrix}, \quad D_2 = \begin{pmatrix} 1 & 0 & 0 \\ 0 & 1+\phi^2 & 0 \\ 0 & 0 & 1 \end{pmatrix}.$$

Although both ϕ_{i21} *are the coefficients of regressing the second measurement on the first, they actually have different interpretations and take different values: For Subject 1, the measurement at time 2 is regressed on that at time 1, but for Subject 2, the measurement at time 3 is regressed on that at time 1. In particular, it is evident that* $\phi_{121} = \phi$*, while* $\phi_{221} = \phi^2$.

Similar results and statements hold for the innovation variances. This difference in the values of the regression parameters or the lack of coherence implies that a naive approach that simply relates the ϕ_{itj} to some covariates may not be statistically prudent when the data are unbalanced.

3.10.2 The Incomplete Data and The EM Algorithm

To deal with the missing data, we develop a "generalized EM algorithm" in the context of the modified Cholesky decomposition and compute the MLE of the parameters. Let $\boldsymbol{Y}_i$ be a $p_i \times 1$ vector containing the responses for subject i, where $i = 1, \ldots, n$. The $\boldsymbol{Y}_i$ are assumed to follow the model

$$\boldsymbol{Y}_i = X_i\beta + e_i,$$

where X_i is a $p_i \times p$ known matrix of covariates, β is a $p \times 1$ vector of unknown regression parameters, and the e_i are independently distributed as $N(0, \boldsymbol{\Sigma}_i)$. We assume that e_i is a sub-vector of a larger $p \times 1$ vector e_i^* that corresponds to the same set of p observation times $t_1, \ldots, t_p$, for all i. This model assumption is valid in a typical setting of longitudinal data when the measurements are collected at the same set of scheduled time points for all subjects, though for a particular subject some measurements might be missing. Under the above model assumptions, $\boldsymbol{\Sigma}_i$ is a sub-matrix of $\boldsymbol{\Sigma}_i^* = \text{var}(e_i^*)$ and using the modified Cholesky decomposition, there exists a unique lower triangular matrix T_i with 1's as main diagonal entries and a unique diagonal matrix D_i with positive diagonal entries such that $T_i\boldsymbol{\Sigma}_i^*T_i' = D_i$. The below-diagonal entries of T_i are the negatives of the autoregressive coefficients, ϕ_{itj}, in $\widehat{e}_{it}^* = \sum_{j=1}^{t-1} \phi_{itj}e_{ij}^*$, the linear least-squares predictor of e_{ij}^* based on its predecessors $e_{i(t-1)}^*, \ldots, e_{i1}^*$. The diagonal entries of D_i are the innovation variances $\sigma_{it}^2 = \text{var}(e_{it}^* - \widehat{e}_{it}^*)$, where $1 \le t \le p$ and $1 \le i \le n$. The parameters ϕ_{itj} and $\log \sigma_{it}^2$ are unconstrained and are modeled as in (3.54). We assume that there is no missing value in the covariates.

To compute the MLE, we use an iterative EM algorithm for the incomplete data model (Dempster et al., 1977). The algorithm consists of two parts. The first part applies the generalized least-squares solution to update β:

$$\tilde{\beta} = \Big(\sum_{i=1}^{n} X_i'\boldsymbol{\Sigma}_i^{-1}X_i\Big)^{-1}\Big(\sum_{i=1}^{n} X_i'\boldsymbol{\Sigma}_i^{-1}\boldsymbol{Y}_i\Big), \tag{3.56}$$

which is obtained by maximizing the likelihood function with respect to β while holding γ and λ fixed at their current values. The second part comprises one iteration of a generalized EM algorithm to update λ and γ, using e_i^* as complete data and sub-vectors $e_i = \boldsymbol{Y}_i - X_i\beta$ as observed data, and setting β equal to its current value. The algorithm iterates between the two parts until convergence. Further details on the EM algorithm for estimating the parameters and the asymptotic inference are given next following the approach of Huang et al. (2012).

The E-step of the generalized EM algorithm relies on the parametrization of the modified Cholesky decomposition of the covariance matrix. Minus twice the log-likelihood function for the complete data, except for a constant, is given by

$$-2l = \sum_{i=1}^{n}(\log|\boldsymbol{\Sigma}_i^*| + e_i^{*'}\boldsymbol{\Sigma}_i^{*-1}e_i^*) = \sum_{i=1}^{n}\{\log|\boldsymbol{\Sigma}_i^*| + \mathrm{tr}(\boldsymbol{\Sigma}_i^{*-1}V_i)\}, \quad (3.57)$$

where $V_i = e_i^* e_i^{*'}$. Let Q be the expected log-likelihood given the observed data and the current parameter values. Denote $\widehat{V}_i = E(e_i^* e_i^{*'}|e_i)$, whose computation is detailed later. Then,

$$-2Q = \sum_{i=1}^{n}\{\log|\boldsymbol{\Sigma}_i^*| + \mathrm{tr}(\boldsymbol{\Sigma}_i^{*-1}\widehat{V}_i)\}. \quad (3.58)$$

We now give two expressions for $-2Q$ that are useful when deriving the required steps of the EM algorithm. Define $\mathrm{RS}_{it} = (e_{it}^* - \sum_{j=1}^{t-1} e_{ij}^* z_{itj}'\gamma)^2$ and $\widehat{\mathrm{RS}}_{it} = E(\mathrm{RS}_{it}|e_i)$. The modified Cholesky decomposition $T_i\boldsymbol{\Sigma}_i^* T_i' = D_i$ can be used to get

$$-2Q = \sum_{i=1}^{n}\sum_{t=1}^{p}\left(\log\sigma_{it}^2 + \frac{\widehat{\mathrm{RS}}_{it}}{\sigma_{it}^2}\right). \quad (3.59)$$

For $t > 1$, denote $Z_{it}' = (z_{it1}, \ldots, z_{it(t-1)})$, and let $\widehat{V}_{itt} = \widehat{V}_i[t, t]$, $\widehat{V}_{it}^{(t-1)} = \widehat{V}_i[1{:}(t-1), t]$, $\widehat{V}_i^{(t-1)} = \widehat{V}_i[1{:}(t-1), 1{:}(t-1)]$ be sub-matrices of $\widehat{V}_i$. We also make the convention that $\widehat{V}_{i1}^{(0)} = 0$ and $\widehat{V}_i^{(0)} = 0$. Using the fact that $\widehat{\mathrm{RS}}_{it}$ is the (t, t)th element of the matrix $T_i\widehat{V}_i T_i'$, we obtain from (3.59) that

$$-2Q = \sum_{i=1}^{n}\sum_{t=1}^{p}\left(\log\sigma_{it}^2 + \frac{\widehat{V}_{itt}}{\sigma_{it}^2}\right) + \sum_{i=1}^{n}\sum_{t=1}^{p}\sigma_{it}^{-2}(-2\gamma' Z_{it}'\widehat{V}_{it}^{(t-1)} + \gamma' Z_{it}'\widehat{V}_i^{(t-1)} Z_{it}\gamma). \quad (3.60)$$

The calculation of $\widehat{V}_i$ is as follows. Note that $\widehat{V}_i = E(e_i^* e_i^{*'}|e_i) = \widehat{e_i^*}\widehat{e_i^*}' + \mathrm{var}(e_i^*|e_i)$ with $\widehat{e_i^*} = E(e_i^*|e_i)$. Write

$$e_i^* = \begin{pmatrix} e_i \\ e_i^+ \end{pmatrix} \sim N(0, \boldsymbol{\Sigma}_i^*), \qquad \boldsymbol{\Sigma}_i^* = \begin{pmatrix} \boldsymbol{\Sigma}_{i11}^* & \boldsymbol{\Sigma}_{i12}^* \\ \boldsymbol{\Sigma}_{i21}^* & \boldsymbol{\Sigma}_{i22}^* \end{pmatrix}.$$

Then, from the standard results for conditional distribution of multivariate normal distributions it follows that

$$E(e_i^*|e_i) = \begin{pmatrix} I \\ \boldsymbol{\Sigma}_{i21}^*\boldsymbol{\Sigma}_{i11}^{*-1} \end{pmatrix} e_i, \quad \mathrm{var}(e_i^*|e_i) = \begin{pmatrix} 0 & 0 \\ 0 & \boldsymbol{\Sigma}_{i22}^* - \boldsymbol{\Sigma}_{i21}^*\boldsymbol{\Sigma}_{i11}^{*-1}\boldsymbol{\Sigma}_{i12}^* \end{pmatrix}. \quad (3.61)$$

Using (3.59) and (3.60), the update of γ and λ proceeds as follows. For fixed λ, since $-2Q$ is a quadratic form in γ, it is minimized by

$$\tilde{\gamma} = \Big(\sum_{i=1}^{n}\sum_{t=1}^{p} \sigma_{it}^{-2} Z_{it}' \widehat{V}_i^{(t-1)} Z_{it}\Big)^{-1} \sum_{i=1}^{n}\sum_{t=1}^{p} \sigma_{it}^{-2} Z_{it}' \widehat{V}_{it}^{(t-1)}. \tag{3.62}$$

For fixed γ, since optimization of $-2Q$ over λ is not simple and does not have a closed-form expression we resort to the Newton–Raphson algorithm. From $\log \sigma_{it}^2 = z_{it}'\lambda$, simple calculation yields

$$\frac{\partial Q}{\partial \lambda} = -\frac{1}{2}\sum_{i=1}^{n}\sum_{t=1}^{p}\Big(1 - \frac{\widehat{\mathrm{RS}}_{it}}{\sigma_{it}^2}\Big) z_{it}$$

and

$$\frac{\partial^2 Q}{\partial \lambda \partial \lambda'} = -\frac{1}{2}\sum_{i=1}^{n}\sum_{t=1}^{p}\frac{\widehat{\mathrm{RS}}_{it}}{\sigma_{it}^2} z_{it} z_{it}'.$$

The Newton–Raphson algorithm updates the current values $\lambda^{(0)}$ to $\lambda^{(1)}$ using

$$\lambda^{(1)} = \lambda^{(0)} + \Delta\lambda, \qquad \Delta\lambda = -\Big(\frac{\partial^2 Q}{\partial \lambda \partial \lambda'}\Big)^{-1}\frac{\partial Q}{\partial \lambda}. \tag{3.63}$$

For the generalized EM algorithm, there is no need to do a full iteration of the Newton–Raphson algorithm. We only need to make sure that $Q(\lambda)$ increases at each iteration, using partial stepping such as step-halving, if necessary. Recall that step-halving works as follows: If $Q(\lambda^{(1)}) \leq Q(\lambda^{(0)})$, we replace $\Delta\lambda$ by its half in the update $\lambda^{(1)} = \lambda^{(0)} + \Delta\lambda$, and continue doing so until $Q(\lambda^{(1)}) > Q(\lambda^{(0)})$.

The steps of the algorithm are summarized as follows:

The Generalized EM Algorithm:

---

1. Initialization: set $\boldsymbol{\Sigma}_i^* = I, i = 1, \ldots, n$.
2. Using the current estimates of γ and λ (or $\boldsymbol{\Sigma}_i^*$ in the first iteration), compute the updated estimate $\tilde{\beta}$ of β using (3.56).
3. Compute $\widehat{V}_i$, $i = 1, \ldots, n$, where the relevant conditional expectations are calculated using (3.61).
4. Using the current estimates of β and λ, update γ using (3.62).
5. Using the current estimates of β and γ, update the current estimate $\lambda^{(0)}$ to $\lambda^{(1)}$ using one step of Newton–Raphson as in (3.63). Use step-halving to guarantee that the criterion is increased.
6. Iterate steps 2–5 until convergence.

Asymptotic Inference: The asymptotic covariance matrix of the parameters is computed after the EM algorithm (Oakes, 1999). The observed information of (γ, λ) evaluated at $(\tilde{\gamma}, \tilde{\lambda})$ can be approximated by

$$\begin{pmatrix} \sum_{i=1}^{p} S(\tilde{\gamma}; \boldsymbol{Y}_i) S^t(\tilde{\gamma}; \boldsymbol{Y}_i) & \sum_{i=1}^{p} S(\tilde{\gamma}; \boldsymbol{Y}_i) S^t(\tilde{\lambda}; \boldsymbol{Y}_i) \\ \sum_{i=1}^{p} S(\tilde{\lambda}; \boldsymbol{Y}_i) S^t(\tilde{\gamma}; \boldsymbol{Y}_i) & \sum_{i=1}^{p} S(\tilde{\lambda}; \boldsymbol{Y}_i) S^t(\tilde{\lambda}; \boldsymbol{Y}_i) \end{pmatrix}, \tag{3.64}$$

where

$$S(\tilde{\gamma}; \boldsymbol{Y}_i) = \left.\frac{\partial Q_i}{\partial \gamma}\right|_{\gamma=\tilde{\gamma}} = \sum_{t=1}^{p} \sigma_{it}^{-2} (Z_{it}' \widehat{V}_{it}^{(t-1)} - Z_{it}' \widehat{V}_i^{(t-1)} Z_{it} \gamma)\Big|_{\gamma=\tilde{\gamma}}$$

and

$$S(\tilde{\lambda}; \boldsymbol{Y}_i) = \left.\frac{\partial Q_i}{\partial \lambda}\right|_{\lambda=\tilde{\lambda}} = -\frac{1}{2} \sum_{t=1}^{p} \left(1 - \frac{\widehat{\mathrm{RS}}_{it}}{\sigma_{it}^2}\right) z_{it}\Big|_{\lambda=\tilde{\lambda}},$$

where Q_i is the term in Q corresponding to subject i. The asymptotic covariance matrix of the MLE $(\widehat{\gamma}, \widehat{\lambda})$ is obtained as the inverse of the observed information matrix (3.64), evaluated at the estimated parameter values. Since $\widehat{\beta}$ and $(\widehat{\gamma}, \widehat{\lambda})$ are asymptotically independent, the asymptotic covariance matrix of $\widehat{\beta}$ is estimated by $(\sum_{i=1}^{p} X_i' \widehat{\Sigma}^{-1} X_i)^{-1}$.

3.11 A DATA EXAMPLE: FRUIT FLY MORTALITY RATE

In this section, we illustrate the GLM methodology for the covariance matrix of an incomplete longitudinal data following Garcia et al. (2012). The fruit fly mortality (FFM) data (Zimmerman and Núñez-Antón 2009, p. 21) consist of age-specific mortality measurements from $n = 112$ cohorts (subjects) of *Drosophila melanogaster* (the common fruit fly). To obtain these mortality measurements, the investigators replicated 56 recombinant inbred lines to get 500 to 1000 fruit flies for each cohort. Each day during the study, the number of dead flies within each cohort were counted and removed. The FFM data consist of these counts pooled into eleven 5-day intervals. To analyze the mortality rates over each cohort i, $i = 1, \ldots, 112$, we let $N_i(t)$ denote the number of flies still alive at the beginning of the tth 5-day interval, where $t = 1, \ldots, 11$. An appropriate measure for the raw mortality rate is $-\log\{N_i(t+1)/N_i(t)\}$, and a log transform of this rate is approximately normally distributed. Then, the response variable in this longitudinal study is $\boldsymbol{Y}_i = (Y_{i,1}, \cdots, Y_{i,11})'$, where each $Y_{i,t} = \log\left[-\log\{N_i(t+1)/N_i(t)\}\right]$ denotes the log mortality rate for cohort i at time t. For unknown reasons, approximately 22%

TABLE 3.2 Summary statistics for the FFM data

Time	1	2	3	4	5	6	7	8	9	10	11
n_o	85	89	94	98	103	104	100	85	73	67	56
$\bar{Y}$	−3.789	−3.409	−3.418	−2.641	−1.996	−1.283	−0.775	−0.481	−0.567	−0.511	−0.408
s^2	0.701	1.084	1.657	2.609	2.167	1.726	1.054	0.630	0.428	0.334	0.417

Note: n_o is the number of observed values at each time point, $\bar{Y}$ is the sample mean, and s^2 is the sample variance of the available data.

of the log mortality rates are missing. The missingness is intermittent in that some missing values are followed by observed values.

According to Zimmerman and Núñez-Antón (2009, pp. 21–23), the study has two main objectives. First, to find a parsimonious model that adequately describes the change in log mortality rates over time, and second, to find the relation between the log mortality rate at any given age (time) with the log mortality rate at a previous age. Of course, these objectives are related to suitable modeling of the mean vector μ and the covariance matrix Σ of the data.

A naive approach to estimating the sample mean and covariance of an incomplete multivariate data is to pretend that there are no missing values and use the standard formulas for the means, variances, and covariances, but skip values or their products in the sums when not available and divide the resulting sums by the number of available terms. This is a simple instance of the available case analysis where the analyst tries not to waste any observed values. The sample means and variances in Table 3.2 were computed using the available cases for each measurement time. The sample correlation matrix is computed using *pairwise* available cases, which in the presence of missing data, will lead to an imbalance in the number of contributed terms to the calculation. This imbalance might cause the available-case covariance matrix **S** to not be positive semidefinite. Fortunately, for the FFM data we do have a positive-definite available-case sample covariance matrix, since its smallest eigenvalue is 0.131.

As expected in most longitudinal studies and as can be seen easily in Table 3.3, the correlations are mostly positive and they become smaller as the time lags between

TABLE 3.3 Sample correlations (below the main diagonal), innovation variances (along the main diagonal), and generalized autoregressive parameters (above the main diagonal) for the FFM data

Time	1	2	3	4	5	6	7	8	9	10	11
1	0.701	**0.722**	0.331	0.059	−0.014	0.165	0.047	−0.141	−0.007	−0.021	0.149
2	**0.585**	0.719	**0.728**	0.258	0.178	0.016	0.069	0.028	0.049	0.171	−0.111
3	**0.531**	**0.711**	0.763	**0.804**	−0.160	0.039	−0.088	0.093	−0.080	−0.301	−0.298
4	**0.476**	**0.621**	**0.776**	0.995	**0.693**	−0.253	−0.040	−0.089	0.085	0.248	0.299
5	**0.340**	**0.525**	**0.591**	**0.787**	0.998	**0.808**	−0.004	−0.020	0.007	−0.140	−0.095
6	**0.323**	**0.383**	**0.432**	**0.533**	**0.821**	1.726	**0.577**	−0.007	−0.213	−0.039	0.085
7	0.206	0.253	0.212	0.277	**0.573**	**0.781**	0.518	**0.564**	0.153	0.120	−0.063
8	0.004	0.111	0.080	0.035	0.308	**0.512**	**0.736**	0.329	**0.447**	0.074	0.183
9	−0.025	0.070	−0.024	0.039	0.053	0.050	**0.399**	**0.592**	0.271	**0.261**	0.089
10	−0.006	0.161	−0.159	0.170	0.011	0.029	0.311	**0.336**	**0.462**	0.197	**0.185**
11	0.088	−0.052	−0.153	0.281	0.177	0.192	0.253	0.289	0.292	**0.373**	0.270

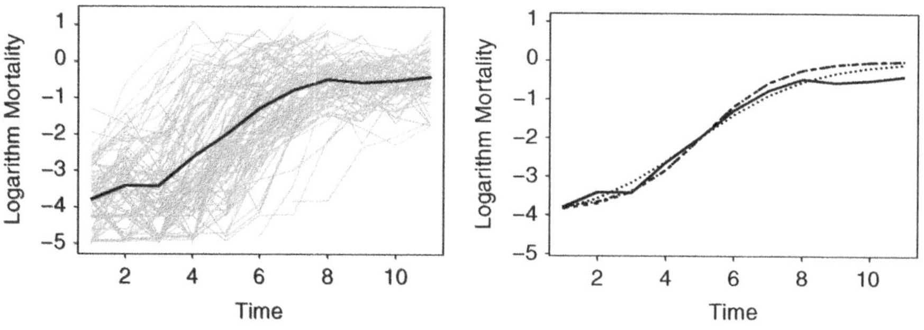

FIGURE 3.1 Profile plot of the FFM (left) with the darker line denoting the mean trend. Plot at right displays the mean trend (solid line) and fitted logistic mean with (i) covariance $\sigma^2 I$ (dotted line); (ii) covariance as in (3.66) and (3.67) (dashed–dotted line); and (iii) covariance as in (3.66) and (3.68) (dashed line).

measurements increase such that measurements closer in time are more correlated. Interestingly, for the FFM data, most of the correlations (appearing in boldface) up to lag 6 are fairly large. Given that the correlations along the subdiagonals are generally first increasing and then decreasing, assuming a stationary covariance model would be inappropriate. Thus, a data-based method of modeling the covariance matrix that properly accounts for these observed facts and that can handle the potential nonstationarity is bound to lead to better inferences for the mean parameters.

For model formulation, we use some graphical tools such as regressograms in search for a meaningful mean–covariance model. First, the profile plot of the FFM available data in Figure 3.1 suggests that the log mortality rates generally increase with time. The overall increasing pattern of the sample mean (indicated by the solid line) seems to follow an "S-shape" or a logistic function of time. Thus, we propose the following nonlinear function of time as a model for the mean:

$$\mu_t(\beta) = \frac{\beta_0}{1 + \exp\{-(t - \beta_1)/\beta_2\}}. \tag{3.65}$$

The three parameters of the model are statistically interpretable, namely, β_0 corresponds to the asymptotic log mortality rate, β_1 is the time point when half the asymptotic value is achieved, and β_2 measures the curvature at this time.

Examining the sample innovariogram in Figure 3.2 suggests a cubic polynomial in time for $\log \sigma_t^2$:

$$\log \sigma_t^2 = \lambda_1 + \lambda_2 t + \lambda_3 t^2 + \lambda_4 t^3 + \epsilon_{t,q}, \tag{3.66}$$

and the regressogram in Figure 3.2 suggests the following two possibly competing models for ϕ_{tj}:

$$\phi_{tj} = \gamma_1 + \gamma_2(t-j) + \gamma_3(t-j)^2 + \gamma_4(t-j)^3 + \epsilon_{t,d}, \tag{3.67}$$

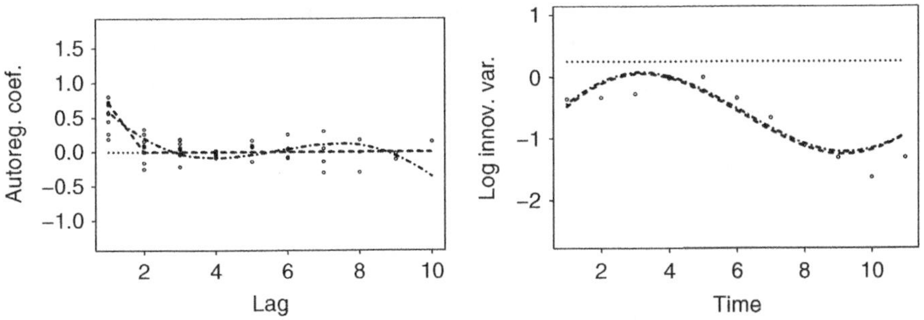

FIGURE 3.2 Sample regressogram (left) and innovariogram (right) for the FFM data (circles) with overlaying fitted mean–covariance models: logistic mean with (i) covariance $\sigma^2 I$ (dotted line); (ii) covariance as in (3.66) and (3.67) (dashed–dotted line); and (iii) covariance as in (3.66) and (3.68) (dashed line).

and

$$\phi_{t,t-1} = \gamma_1, \phi_{t,j} = 0, 2 \leq (t - j) \leq t, \tag{3.68}$$

where the error terms $\epsilon_{t,q}$ and $\epsilon_{t,d}$ have zero mean and unknown finite variances. As far as parsimony is concerned, for the FFM data with 11 repeated measurements on each cohort, fitting the saturated covariance matrix requires 66 possibly distinct and constrained parameters. However, our data-driven procedure for formulating models (3.66) and (3.67) for the (T, D)-pair uses only $4 + 4 = 8$ unconstrained parameters, while those based on (3.66) and (3.68) require $4 + 1 = 5$ parameters. Of course, this is a considerable gain in terms of parsimony and ease of estimation due to the unconstrained nature of the parameters of these models. We compare the performance of these proposed mean–covariance models using the log-likelihood and the BIC (Pan and MacKenzie, 2003):

$$-2\widehat{L}_{\max}(\widehat{\beta}', \widehat{\lambda}', \widehat{\gamma}')/n + (p + q + d)\log(n)/n, \tag{3.69}$$

where $\widehat{L}_{\max}$ is the maximized log-likelihood evaluated at $\widehat{\beta}, \widehat{\lambda}, \widehat{\gamma}$ which have dimensions $p \times 1$, $q \times 1$, and $d \times 1$, respectively.

From Figure 3.1, it is evident that the fitted logistic mean model for all three covariance models captures the overall sample mean pattern well especially up to time $t = 8$. After $t = 8$, the fitted means tend to deviate greatly which could be a consequence of the difficulty in estimating parameters at time points with a larger percentage of missing values (Garcia et al. 2012).

From Table 3.4 and Figure 3.2, it is evident that the naive model which completely ignores the strong correlations present in the data gives the worst fit. The second and third models improve over the first in accounting for the correlation structure in the data and both have smaller standard errors for the estimates of (β, λ, γ) (see

TABLE 3.4 Maximum log-likelihood ($\widehat{L}_{\max}$) and BIC for logistic mean model and different covariance models for the FFM data

Covariance Model	$\hat{L}_{max}$	BIC
$\sigma^2 I$	−1471.839	26.451
(3.66) and (3.67)	−261.894	5.140
(3.66) and (3.68)	−242.334	4.664

Table 3.5). Regardless of the posited model for the $\phi_{t,j}$, the resulting parameter estimates of λ in model (3.66) are similar and capture the behavior of the innovariogram well. The third fitted model (3.66) and (3.68) has maximum log-likelihood, lowest BIC, and smallest standard errors for the covariance parameter λ. Furthermore, Figure 3.2 indicates that model (3.68) captures the overall pattern of the sample regressogram better than does model (3.67). The results from Tables 3.4, 3.5, and Figure 3.2 imply that the third covariance model is the best among the three considered.

In summary, our analysis suggests that the log mortality rate for the FFM data is best modeled using a logistic model (3.65) for the mean μ, and models (3.66) and (3.68) for the (T, D)-pair of the covariance matrix.

Finally, we address the second objective of the FFM study which was to model the relationship between the log mortality rate at adjacent time points. From (3.25) and (3.68) a plausible model is the AR(1) model:

$$y_t - \mu_t = \phi(y_{t-1} - \mu_{t-1}) + \epsilon_{t,n}, \tag{3.70}$$

with estimated $\widehat{\phi} = \widehat{\gamma}_1 = 0.717$ (see Table 3.5).

In the search for a parsimonious model for the log mortality rate or reasonable models for the mean μ and covariance matrix Σ, we model the mean as the logistic

TABLE 3.5 Estimated parameters (est) and estimated standard errors ($\widehat{\text{se}}$) for logistic mean model and three different covariance models for the FFM data.

	$\hat{\beta}_0$	$\hat{\beta}_1$	$\hat{\beta}_2$	$\hat{\sigma}$	$\hat{\gamma}_1$	$\hat{\gamma}_2$	$\hat{\gamma}_3$	$\hat{\gamma}_4$	$\hat{\lambda}_1$	$\hat{\lambda}_2$	$\hat{\lambda}_3$	$\hat{\lambda}_4$
					Covariance Model: $\sigma^2 I$							
est	−4.216	4.829	−1.670	1.134								
$\widehat{\text{se}}$	0.190	0.212	0.133									
					Covariance Model: (3.66) and (3.67)							
est	−3.866	5.102	−1.088		0.047	−0.492	0.141	−0.528	−0.575	−1.297	−0.203	0.851
$\widehat{\text{se}}$	0.061	0.096	0.058		0.015	0.071	0.054	0.053	0.045	0.188	0.170	0.177
					Covariance Model: (3.66) and (3.68)							
est	−3.922	5.065	−1.104		0.717				−0.607	−1.288	−0.188	0.884
$\widehat{\text{se}}$	0.069	0.101	0.062		0.020				0.042	0.162	0.152	0.157

function (3.65), and model the covariance with three different models: a simple, yet naive diagonal covariance $\sigma^2 I$; a covariance model as in (3.66) and (3.67); and a third as in (3.66) and (3.68). The fits for each model are assessed using the log-likelihood and BIC, available in Table 3.4, and resulting parameter estimates in Table 3.5 (Garcia et al. 2012).

3.12 SIMULATING RANDOM CORRELATION MATRICES

Simulating values of a random variable is a well-understood subject. How does one simulate from a random correlation matrix? Simulating random or "typical" correlation matrices are of importance in various areas of statistics (Joe, 2006), engineering and signal processing (Holmes, 1991), finance, and numerical analysis.

The main obstacles encountered are the positive-definiteness constraint on a correlation matrix R and that all its diagonal entries are the same and equal to one. Although there are techniques based on the spectral and Cholesky decompositions for handling the positive-definiteness requirement, they do not apply directly when there are additional constraints such as stationarity or constancy along certain (sub)diagonals (Pourahmadi, 1999, Section 2.6) or zero entries.

An approach due to Joe (2006) reparameterizes a constrained $p \times p$ correlation matrix $R = (\rho_{ij})$ using the correlations $\rho_{i,i+1}$ for $i = 1, \cdots, p-1$ and the partial correlations $\rho_{ij|i+1,\cdots,j-1}$ for $j - i \geq 2$. This amounts to swapping the correlation matrix R by a *partial autocorrelation matrix* Π, where the latter is simply a symmetric $p \times p$ matrix with 1's on the main diagonal and off-diagonal elements vary freely in the interval $(-1, 1)$. Such a not necessarily positive-definite matrix is also known as a *proto correlation matrix* (Kurowicka and Cooke, 2003). We highlight the distinguished roles of these partial autocorrelations in providing a one-to-one correspondence between R and Π which leads to unconstrained and interpretable reparameterizations of correlation matrices.

Full partial correlations (of two variables given the *rest*) have been around and used for over a century in the context of multiple regression and multivariate statistics (Anderson, 2003). The more specialized partial autocorrelation (of two variables given the *intervening* variables) have proved to be indispensable in the study of stationary processes and situations dealing with Toeplitz matrices such as Szegö's orthogonal polynomials, trigonometric moment problems, geophysics, digital signal processing and filtering, inverse scattering, interpolation theory of functions and operators (Ramsey, 1974; Pourahmadi, 2001; Simon, 2005), identification of ARMA models, the maximum likelihood estimation of their parameters (Jones, 1980), and simulating a random or "typical" ARMA model (Jones, 1987).

Starting with the variance–correlation decomposition of a general $\Sigma = DRD$, we focus on reparameterizing its correlation matrix $R = (\rho_{ij})$ in terms of entries of a simpler symmetric matrix $\Pi = (\pi_{ij})$ where $\pi_{ii} = 1$ and for $i < j$, π_{ij} is the *partial autocorrelation* between Y_i and Y_j adjusted for the *intervening* (not the remaining) variables. More precisely, following Joe (2006) for $j = 1, \cdots, p-k, k = 1, \cdots, p-1$, let $r_1'(j,k) = (\rho_{j,j+1}, \ldots, \rho_{j,j+k-1}), r_3'(j,k) = (\rho_{j+k,j+1}, \ldots, \rho_{j+k,j+k-1})$, and

$R_2(j,k)$ is the correlation matrix corresponding to components $(j+1,\ldots,j+k-1)$. Then, the partial autocorrelations between Y_j and Y_{j+k} adjusted for the intervening variables, denoted by $\pi_{j,j+k} \equiv \rho_{j,j+k|j+1,\ldots,j+k-1}$, are computed using the expression

$$\pi_{j,j+k} = \frac{\rho_{j,j+k} - r_1'(j,k)R_2(j,k)^{-1}r_3(j,k)}{[1 - r_1'(j,k)R_2(j,k)^{-1}r_1(j,k)]^{1/2}[1 - r_3'(j,k)R_2(j,k)^{-1}r_3(j,k)]^{1/2}}. \tag{3.71}$$

In what follows and in analogy with R, it is convenient to arrange these partial autocorrelations in a matrix Π where its $(j, j+k)$th entry $\pi_{j,j+k}$ is defined as in (3.71). Note that the link function $g(\cdot)$ in (3.71) that maps a correlation matrix R into the partial autocorrelation matrix Π is indeed invertible, so that solving (3.71) for $\rho_{j,j+k}$ one obtains

$$\rho_{j,j+k} = r_1'(j,k)R_2(j,k)^{-1}r_3(j,k) + D_{jk}\,\pi_{j,j+k}, \tag{3.72}$$

where D_{jk} is the denominator of the expression in (3.71). Note that the set of $p(p-1)/2$ partial correlations or the matrix Π depends on the order or labeling of the variables in Y. Thus, it is not unique unless an order is imposed on the variables. Of course, such order exists naturally for longitudinal data; for other settings we assume an order is fixed. Then, the above formulae clearly establish a one-to-one correspondence between the matrices R and Π. A major advantage of parameterization in terms of the partial autocorrelations is that these can take any value in $(-1, 1)$ regardless of the values of the others. This is in sharp contrast to the entries of the correlation matrix R, where the off-diagonal elements take values in $(-1, 1)$, but they are constrained so that R is positive definite. Note that unlike R, and the matrix of full partial correlations (ρ^{ij}) constructed from $\mathbf{\Sigma}^{-1}$, Π has a simpler structure in that its entries are free to vary in the interval $(-1, 1)$. If needed, using the Fisher z-transform Π can be mapped to the matrix $\tilde{\Pi}$ where its off-diagonal entries take values in the entire real line $(-\infty, +\infty)$.

Compared to the long history of using the PACF in time series analysis (Quenouille, 1949), research on establishing a one-to-one correspondence between a general covariance matrix and (D, Π) has a rather short history. An early work in the Bayesian context is due to Eaves and Chang (1992), followed by Zimmerman (2000) for longitudinal data, Dégerine and Lambert-Lacroix (2003) for the nonstationary time series, and Kurowicka and Cooke (2003), and Joe (2006) for a general random vector. The fundamental determinantal identity:

$$|\Sigma| = \left(\prod_{i=1}^{p} \sigma_{ii}\right) \prod_{i=2}^{p} \prod_{j=1}^{i-j} \left(1 - \pi_{ij}^2\right), \tag{3.73}$$

has been redicovered recently by Dégerine and Lambert-Lacroix (2003), Kurowicka and Cooke (2003) and Joe (2006), but its origin can be traced to a notable and somewhat neglected paper of Yule (1907, equ. 25).

The identity (3.73) plays a central role in Joe's (2006) method of generating random correlation matrices whose distributions are *independent of the order of variables* in Y, and in Daniels and Pourahmadi's (2009) introduction of priors for the Bayesian analysis of correlation matrices. These papers employ independent linearly transformed Beta priors on $(-1, 1)$ for the partial autocorrelations π_{ij}. However, Jones (1987) seems to be the first to use such Beta priors in simulating data from "typical" ARMA models.

3.13 BAYESIAN ANALYSIS OF COVARIANCE MATRICES

In this section, a primer on Bayesian analysis of covariance matrices is presented. The subject is vast and growing; however, in the regression-based approach to covariance estimation we focus on priors for a covariance matrix introduced via priors for the ensuing regression parameters (Daniels and Pourahmadi, 2002; Smith and Kohn, 2002; Fox and Dunson, 2011). Furthermore, it is instructive to note that in a regularized regression setup, exponentiating the penalty function $p(\beta)$ as in $ce^{-p(\beta)}$ leads to a bona fide prior for the regression parameters (Tibshirani, 1996).

We review the more classic trends in introducing priors for covariance matrices and provide some key references as a starting point for further research in comparing the classic and modern ways of introducing priors. Traditionally, in Bayesian approaches to inference for Σ the Jefferys' improper prior and the conjugate inverse Wishart (IW) priors are used. For reviews of the earlier work in this direction, see Lin and Perlman (1985) and Brown et al. (1994).

The success of Bayesian computation and Markov Chain Monte Carlo (MCMC) in the late 1980s did open up the possibility of using more flexible and elaborate nonconjugate priors for covariance matrices (see Yang and Berger, 1994; Daniels and Kass, 2001; Wong et al., 2003; Hoff, 2009). Some of these priors were constructed and inspired by certain useful and desirable features of IW and have led to the generalized inverse Wishart (GIW) priors introduced by Brown et al. (1994), Daniels and Pourahmadi (2002), Smith and Kohn (2002), Barnard et al. (2000), Wong et al. (2003), which rely either on the Cholesky or the variance–correlation decompositions. Our brief review of the progress in Bayesian covariance estimation is in a somewhat chronological order starting with priors put on the components of the spectral decomposition.

Priors on the Spectral Decomposition: Starting with the remarkable work of Stein (1956), efforts to improve estimation of a covariance matrix have been confined mostly to shrinking the eigenvalues of the sample covariance matrix toward a common value (Dey and Srinivasan, 1985; Lin and Perlman, 1985; Haff, 1991; Yang and Berger, 1994; Daniels and Kass, 1999; Hoff, 2009). Such covariance estimators have lower risk than the sample covariance matrix. Intuitively, shrinking the eigenvectors is expected to further improve or reduce the estimation risk (Daniels and Kass, 1999, 2000; Johnstone and Lu, 2009).

There are three broad classes of priors which are based on unconstrained parameterizations of a covariance matrix using its spectral decomposition. These have the goal of shrinking some functions of the off-diagonal entries of $\mathbf{\Sigma}$ or the corresponding correlation matrix towards a common value like zero. Consequently, estimation of the $p(p-1)/2$ dependence parameters is reduced to that of estimating a few parameters.

The log matrix prior due to Leonard and Hsu (1992) is based on the matricial logarithm defined in Section 3.4. It formally places a multivariate normal prior on the entries of $\log \mathbf{\Sigma}$ which leads to introducing a large number of hyperparameters. The flexibility of this class of priors is evident for the covariance matrix of a multivariate normal distribution, yielding much more general hierarchical and empirical Bayes smoothing and inference, when compared with a conjugate analysis involving an IW prior. The prior is not conditionally conjugate, and according to Brown et al. (1994) its major drawback is the lack of statistical interpretability of the entries of $\log \mathbf{\Sigma}$ and their complicated relations to those of Σ as seen in Section 3.4. Consequently, prior elicitation from experts and substantive knowledge cannot be used effectively in arriving at priors for the entries of $\log \mathbf{\Sigma}$ and their hyperparameters.

The reference (noninformative) prior for a covariance matrix in Yang and Berger (1994) is of the form

$$p(\mathbf{\Sigma}) = c[|\mathbf{\Sigma}| \prod_{i<j} (\lambda_i - \lambda_j)]^{-1},$$

where $\lambda_1 > \lambda_2 > \cdots > \lambda_p$ are the ordered eigenvalues of $\mathbf{\Sigma}$ and c is a constant. It is known (Yang and Berger, 1994, p. 1199) that compared to the Jeffreys prior, the reference prior puts considerably more mass near the region of equality of the eigenvalues. Therefore, it is plausible that the reference prior would produce covariance estimators with better eigenstructure shrinkage. It is interesting to note that the reference priors for $\mathbf{\Sigma}^{-1}$ and the eigenvalues of the covariance matrix are the same as $p(\mathbf{\Sigma})$. Expression for the Bayes estimator of the covariance matrix using this prior involve computation of high-dimensional posterior expectations, the computation is done using the hit-and-run sampler in an MCMC setup. An alternative noninformative reference prior for $\mathbf{\Sigma}$ (and the precision matrix) which allows for closed-form posterior estimation is given in Rajaratnam et al. (2008).

It is known (Daniels, 2005) that the Yang and Berger (1994) reference prior implies a uniform prior on the orthogonal matrix P and flat improper priors on the logarithm of the eigenvalues of the covariance matrix. The shrinkage priors of Daniels and Kass (1999) also rely on the spectral decomposition of the covariance matrix, but are designed to shrink the eigenvectors by reparameterizing the orthogonal matrix in terms of $p(p-1)/2$ Givens angles (Golub and Van Loan, 1996) θ between pairs of the columns of the orthogonal matrix P. Since θ is restricted to lie in the interval $(-\pi/2, \pi/2)$, a logit transform will make it unconstrained, and one may put a mean-zero normal prior on them. The statistical relevance and interpretation of the Givens angles are not well-understood at this time. The idea of introducing matrix Bingham distributions as priors on the group of orthogonal matrices (Hoff, 2009) is a major recent contribution to shrinking the eigenvectors of the sample covariance matrix.

Using simulation experiments, Yang and Berger (1994) have compared the performance of their reference prior Bayes covariance estimator to the covariance estimators of Haff (1991) and found it to be quite competitive based on the risks corresponding to the loss functions $L_i, i = 1, 2$. Daniels and Kass (2001) also using simulations have compared the performance of their shrinkage estimator to several other Bayes estimators of covariance matrices, using only the risk corresponding to the L_1 loss function. Their finding is that the Bayes estimators from the Yang and Berger (1994) reference prior does as well as those from Givens-angle prior for some nondiagonal and ill-conditioned matrices, but suffers when the true matrix is diagonal and poorly conditioned.

The Generalized Inverse Wishart Priors: The use of Cholesky decomposition of a covariance matrix or the regression dissection of the associated random vector has a long history and can be traced at least to the work of Bartlett (1933) and Liu (1993). Though the ensuing parameters have nice statistical interpretation, they are not permutation-invariant. It is shown by Brown et al. (1994) that a regression dissection of the inverse Wishart (IW) distribution reveals some of its noteworthy features making it possible to define flexible generalized inverted Wishart (GIW) priors for general covariance matrices.

These priors are constructed by first partitioning a multivariate normal random vector Y with mean zero and covariance matrix $\mathbf{\Sigma}$ into $k \leq p$ subvectors: $Y = (Z_1, \cdots, Z_k)'$, and writing its joint density as the product of a sequence of conditionals:

$$f(y) = f(z_1)f(z_2|z_1)\cdots f(z_k|z_{k-1}, \cdots, z_1).$$

Now, in each conditional distribution one places normal prior distributions on the regression coefficients and inverse Wishart on the prediction variances. The hyperparameters can be structured so as to maintain the conjugacy of the resulting priors. It is known (Daniels and Pourahmadi, 2002; Rajaratnam et al., 2008) that such priors offer considerable flexibility as there are many parameters to control the variability in contrast to the one parameter for IW.

The GIW prior was further refined in Daniels and Pourahmadi (2002) using the finest partition of Y, that is, using $k = p$. In this case, all restrictions on the hyperparameters are removed from the normal and inverse Wishart (gamma) distributions and the prior remains conditionally conjugate, in the sense that the full-conditional of the regression coefficients is normal given the prediction variances, and the full-conditional of prediction variances is inverse gamma given the regression coefficients. For a review of certain advantages of this approach in the context of longitudinal data and some examples of analysis of such data, see Daniels (2005).

Priors on Correlation Matrices: The first use of variance–correlation factorization in Bayesian covariance estimation is due to Barnard et al. (2000) who using the factorization $p(\mathbf{\Sigma}) = p(D, R) = p(D)p(R|D)$ introduced independent priors for the standard deviations in D and the correlations in R. Specifically, they used log normal priors on variances independent of a prior on the whole matrix R capable of inducing uniform $(-1, 1)$ priors on its entries ρ_{ij}. This is done by first deriving the marginal distribution of R when $\mathbf{\Sigma}$ has a standard IW distribution, $W_p^{-1}(I, \nu)$, $\nu \geq p$, with the density

$$f_p(\mathbf{\Sigma}|\nu) = c|\mathbf{\Sigma}|^{-\frac{1}{2}(\nu+p+1)} \exp\left(-\frac{1}{2}tr\mathbf{\Sigma}^{-1}\right).$$

It turns out that

$$f_p(R|\nu) = c|R|^{\frac{1}{2}(\nu-1)(p-1)-1} \prod_{i=1}^{p} |R_{ii}|^{-\nu/2},$$

where R_{ii} is the principal submatrix of R, obtained by deleting its ith row and column. Then, using the marginalization property of the IW (i.e., a principal submatrix of an IW is still an IW), the marginal distribution of each $\rho_{ij}(i \neq j)$ is obtained as

$$f(\rho_{ij}|\nu) = c(1-\rho_{ij}^2)^{\frac{\nu-p-1}{2}}, \ |\rho_{ij}| \leq 1,$$

which is a Beta $\left(\frac{\nu-p+1}{2}, \frac{\nu-p+1}{2}\right)$ on $(-1, 1)$, and reduces to the uniform distribution when $\nu = p+1$. Moreover, by choosing either $p \leq \nu < p+1$ or $\nu > p+1$, one can control the tail of $f(\rho_{ij}|\nu)$, that is, making it heavier or lighter than the uniform. Thus, the above family of priors for R is indexed by a single "tuning" parameter ν.

PROBLEMS

1. Let $\boldsymbol{\Sigma}$ be the covariance matrix of an AR(1) model.

(a) Show that for any $k \geq 2$ the determinant of its $k \times k$ leading principal minor is given by

$$\sigma^{2k}(1-\rho^2)^{k-1}.$$

(b) Show that $\boldsymbol{\Sigma}$ is positive definite, if and only if $-1 < \rho < 1$.

2. Let $\Gamma = (\gamma_{|k-l|})$ be a $p \times p$ *Toeplitz* matrix. Note that it is symmetric with constant (sub)diagonal entries.

(a) Show that

$$\Gamma = \gamma_0 I + \sum_{k=1}^{p} \gamma_k T_k,$$

and T_k is a symmetric matrix with kth superdiagonal and subdiagonal equal to 1 and 0 elsewhere.

(b) Show that Γ is positive definite, if and only if the polynomial

$$P(\theta) = \gamma_0 + \sum_{k=1}^{p} \gamma_k \cos(k\theta), \theta \in R,$$

is nonnegative for all $\theta \in R$.

3. Let Σ be a positive-definite matrix partitioned as

$$\Sigma = \begin{pmatrix} \Sigma_{11} & \Sigma_{12} \\ \Sigma_{21} & \sigma \end{pmatrix}$$

where $\sigma > 0$. Partition C, the Cholesky factor of Σ, as

$$C = \begin{pmatrix} C_1 & 0 \\ C_2 & c \end{pmatrix}.$$

Show that C_1 is the Cholesky factor of Σ_{11} and

$$c = (\sigma - \Sigma_{21}\Sigma_{11}^{-1}\Sigma_{12})^{1/2},\ C_2 = \Sigma_{21}(C_1')^{-1}.$$

4. Consider the partitioned matrix

$$A = \begin{pmatrix} B & C \\ -C & B \end{pmatrix}$$

Show that A is symmetric if and only if $B = B'$ and $C = -C'$. What are the necessary and sufficient conditions on B, C for the matrix A to be positive definite?

5. For the 3×3 matrix in Example 28,

(a) find its eigenvalues, eigenvectors, and the spectral decomposition,

(b) is the matrix positive definite? Invertible?

(c) find its inverse, symmetric square-root, and logarithm.

6. Let L be a $p \times p$ lower triangular matrix with entries

$$L_{11} = 1, L_{j1} = \cos\theta_{j1},\ j = 2, \ldots, p,$$

and

$$L_{jk} = \begin{cases} (\cos\theta_{jk}) \prod_{l=1}^{k-1} \sin\theta_{jl} & k = 2, \ldots, j-1; j = 3, \ldots, p, \\ \prod_{l=1}^{k-1} \sin\theta_{jl} & k = j; j = 2, \ldots, p, \end{cases}$$

where θ_{jk}'s are some real numbers.

(a) Using the basic trigonometric identities show that the diagonal entries of the matrix LL' are equal to 1.

(b) Show that the matrix LL' is a correlation matrix.

(c) Discuss the possibility of using the above factorization to develop GLMs for correlation matrices involving covariates.

7. (a) Let the compound symmetry correlation matrix $R = (1 - \rho)I + \rho \mathbf{1}_p \mathbf{1}_p'$ be factorized as LL' with L a lower triangular matrix. Show that the entries of L are given by:

$$L_{11} = 1,\ L_{jj} = \left[1 - \frac{(j-1)\rho^2}{1+(j-2)\rho}\right]^{1/2},\ j \geq 2,$$

$$L_{ij} = \rho[1 + (j-2)\rho]^{-1} L_{jj}.$$

(b) For the AR(1) correlation matrix $R = (\rho^{|i-j|})$ factorized as above, show that the entries of L are given by

$$L_{i1} = \rho^{|i-1|},\ i = 1, \ldots, p;\ L_{ij} = \frac{\rho^{|i-j|}}{1-\rho^2},\ 2 \leq j < i \leq p.$$

(c) Find the entries of the two factors of the modified Cholesky decomposition of the one-dependent covariance structure given in Example 22.

PART II

COVARIANCE ESTIMATION: REGULARIZATION

CHAPTER 4

REGULARIZING THE EIGENSTRUCTURE

In recent years, various classes of improved high-dimensional covariance estimators have been proposed by assuming *structural* knowledge about $\mathbf{\Sigma}$ such as sparsity, graphical and factor models, and order among the underlying variables. Historically, in the absence of such structural information the focus had been on *rotation equivariant* covariance estimators. In terms of the spectral decomposition of a covariance matrix, an estimator is rotation-equivariant if and only if its eigenvectors are the same as those of the sample covariance matrix. Thus, the differences between two such estimators appear only in their eigenvalues.

In this chapter, we focus first on shrinking the eigenvalues and later regularize the eigenvectors of the sample covariance matrix. There is a long history of shrinking the eigenvalues going back to Stein (1956), but regularizing the eigenvectors is of more recent vintage. The motivations for the latter are computation, interpretability, and consistency of the PCA as a statistical procedure. In fact, the same motivations have inspired regularization of covariance (precision) matrices in other areas of classical multivariate analysis such as clustering, classification, Gaussian graphical models, and multivariate regression. This is one of the most developing and active areas of research in the high-dimensional data setting.

In Section 4.1, we present a broad class of well-conditioned shrinkage/ridge-type estimators based on linear combinations of the sample covariance matrix and the identity matrix of the form

$$\widehat{\mathbf{\Sigma}} = \alpha_1 I + \alpha_2 \mathbf{S}, \tag{4.1}$$

High-Dimensional Covariance Estimation, First Edition. Mohsen Pourahmadi.

where α_1 and α_2 are chosen to minimize a risk function. For the risk corresponding to Frobenius norm, finding the optimal $\alpha_i, i = 1, 2$, amounts to estimating only four functional of the high-dimensional covariance matrix $\mathbf{\Sigma}$ such as its trace and deviation from the sample covariance matrix (Ledoit and Wolf, 2004). These improved estimators of the sample covariance matrix shrink only the eigenvalues of $\mathbf{S}$ toward a central value (Section 2.3) but they leave intact the eigenvectors of the sample covariance matrix. Nevertheless, the methodology is used extensively in various application areas. Establishing consistency of the estimators of the type given in (4.1) brings to fore an interesting interplay among the sample size, dimension, moments, and the dependence structure of the data matrix which is prototypical of the asymptotic theory for high-dimensional covariance estimation.

Improved estimators of the eigenvectors is directly related to the central topic of PCA. Here, *sparsity* of the eigenvectors presents itself as a natural structural assumption which is useful both from the theoretical and application points of view. Interpretation of PCs is important when the variables have physical meanings as in, say, microarray data where each variable corresponds to a specific gene or in neuroimaging where each variable corresponds to a voxel (location). However, since PCs are linear combinations of all the original variables with small, nonzero loadings, it is difficult to interpret them especially when p is large.

Various ad hoc and formal proposals to shrink some of the loadings to zero and hence leading to interpretable PCs are reviewed in Section 4.2. A bona fide search for sparse eigenvectors starts first with the assumption of sparsity of the leading eigenvectors of $\mathbf{\Sigma}$ and then regularizing the leading eigenvectors of the sample covariance matrix. Recognizing that the normalized eigenvectors of $\boldsymbol{X}'\boldsymbol{X}$ and the normalized right singular vectors of the data matrix $\boldsymbol{X}$ are the same, or that PCA and the SVD are closely related to each other, in Section 4.3 we present regression-based methods for sparse PCA. Some of the details of implementing sparse PCA such as defining the percentage of variance explained by the PCs and choosing the tuning parameter are illustrated through a real data example in Section 4.4. Sparse singular value decomposition (SSVD) and inconsistency of PCA in high dimensions are presented in Sections 4.5 and 4.6, respectively.

4.1 SHRINKING THE EIGENVALUES

In this section, we present a shrinkage estimator as a convex linear combination of the sample covariance matrix and a scalar multiple of the identity matrix. One can motivate such estimators by recalling that the sample covariance matrix $\mathbf{S}$ is unbiased for $\mathbf{\Sigma}$, but unstable with considerable risk when $n \ll p$, and a structured covariance estimator like the identity matrix has very little variability but can be severely biased when the structure is misspecified. A natural compromise between these two extremes is a linear combination of the form of (4.1) proposed by Ledoit and Wolf (2004). Thus, one can view a shrinkage estimator as a weighted average of a variance term and a bias term where the weights are chosen to optimize the bias–variance tradeoff. Conditions for consistency of such a simple estimator are given, which reveal the delicate interplay among n, p, moments and dependence structure of the data. A few other estimators involving penalties on the eigenvalues are reviewed at the end of the section.

The coefficients α_1 and α_2 are chosen by minimizing the risk corresponding to the loss function which is a slight modification of the Frobenius norm:

$$L_3(\widehat{\mathbf{\Sigma}}, \mathbf{\Sigma}) = p^{-1}\|\widehat{\mathbf{\Sigma}} - \mathbf{\Sigma}\|^2 = p^{-1}\text{tr}(\widehat{\mathbf{\Sigma}} - \mathbf{\Sigma})^2,$$

defined through the inner product $<A, B>= p^{-1}tr(A'B)$. Note that dividing a norm by the dimension p has the advantage that the norm of the identity matrix is always equal to 1 regardless of the size of the dimension. Also, the loss L_3 does not require matrix inversion which is ideal since $\mathbf{S}$ is not invertible when $n \ll p$.

It follows from basic calculus (Ledoit and Wolf, 2004) that the optimal choice of α_1 and α_2 depends only on the following four scalar functions of the true (but unknown) covariance matrix $\mathbf{\Sigma}$:

$$\mu =< \mathbf{\Sigma}, I >, \alpha^2 = \|\mathbf{\Sigma} - \mu I\|^2, \beta^2 = E\|\mathbf{S} - \mathbf{\Sigma}\|^2, \delta^2 = E\|\mathbf{S} - \mu I\|^2, \tag{4.2}$$

where

$$\alpha^2 + \beta^2 = \delta^2.$$

More precisely, we have the following attractive result:

Lemma 6 *The solution of the minimization problem*

$$min_{\alpha_1,\alpha_2} E||\widehat{\mathbf{\Sigma}} - \mathbf{\Sigma}||^2,$$

for nonrandom α_1, α_2 *leads to:*

$$\widehat{\mathbf{\Sigma}} = \frac{\beta^2}{\delta^2}\mu I + \frac{\alpha^2}{\delta^2}\mathbf{S},$$

and

$$E||\widehat{\mathbf{\Sigma}} - \mathbf{\Sigma}||^2 = \frac{\alpha^2\beta^2}{\delta^2}.$$

Note that $\widehat{\mathbf{\Sigma}}$ is a convex combination of μI and $\mathbf{S}$. In its present form, $\widehat{\mathbf{\Sigma}}$ is not a bona fide *estimator*, because it involves scalars that depend on the unknown $\mathbf{\Sigma}$. Interestingly, once consistent estimators for these scalars are found, substitution in $\widehat{\mathbf{\Sigma}}$ results in a positive-definite estimator of $\mathbf{\Sigma}$. How does one estimate the four parameters in (4.2)? A natural idea is to replace the population covariance matrix $\mathbf{\Sigma}$ by its sample counterpart $\mathbf{S}_n$ and then study the statistical properties of the ensuing estimators. In the discussion of consistency of the estimators, the subscript n is used explicitly for the estimators to emphasize the dependence of the relevant quantities on the sample size.

Theorem 6 *Under Assumptions 1–3 stated later, the estimator*

$$\widehat{\mathbf{\Sigma}}^* = \frac{b_n^2}{d_n^2} m_n I_n + \frac{a_n^2}{d_n^2} \mathbf{S}_n, \tag{4.3}$$

is consistent for $\mathbf{\Sigma}$*, where*

(a) $m_n =< \mathbf{S}_n, I_n >$ *is a consistent estimator of* μ,
(b) $d_n^2 = ||\mathbf{S}_n - m_n I_n||^2$ *is a consistent estimator of* δ^2,
(c) $b_n^2 = min(d_n^2, \bar{b}_n^2)$ *is a consistent estimator of* β^2*, where*

$$\bar{b}_n^2 = 1/n^2 \sum_{k=1}^{n} ||\boldsymbol{X}_{k\cdot}' \boldsymbol{X}_{k\cdot} - \mathbf{S}_n||^2,$$

and $\boldsymbol{X}_{k\cdot}$ *is the kth row of the data matrix* $\boldsymbol{X}$.

We note that given an $n \times p_n$ data matrix $\boldsymbol{X}_n$, computing a consistent estimator $\widehat{\mathbf{\Sigma}}^*$ is rather easy. The required assumptions for consistency of the above estimator is motivated by asking the question: can one estimate consistently the $p_n(p_n + 1)/2$ parameters of $\mathbf{\Sigma}_n$ from the np_n data points in $\boldsymbol{X}_n$? Of course, the answer is in the negative if the former is much larger than the latter. However, for an affirmative answer, one must have at least:

$$\frac{p_n(p_n + 1)}{2} \ll np_n,$$

which leads to:

Assumption 1 There exists a constant K_1 independent of n, such that $p_n/n \leq K_1$.

It says that the dimension p_n should remain fixed or go to infinity with n, but not *too fast*. Existence of the limit of the ratio p_n/n is not required. This should be compared with the ratio $\log(p_n)/n$ which appears in establishing consistency of various covariance estimators obtained through banding and thresholding and under the sparsity assumption.

Consistency can be established by studying the limit of β_n for n large. Note that for β_n to be finite, at least the fourth moment of the random variables in $\boldsymbol{X}$ must exist. It turns out that it is much easier to formulate such moment conditions in terms of the PCs of the data matrix $\boldsymbol{X}$ represented by

$$Y = \boldsymbol{X} V,$$

where the columns of V are the normalized eigenvectors of $\mathbf{\Sigma}$. Recall that with this transformation the rows of Y are i.i.d. p_n-vectors with uncorrelated entries. For simplicity, the ith entry of a generic row is denoted by Y_i.

Assumption 2 There exists a constant K_2 independent of n, such that

$$\frac{1}{p_n}\sum_{i=1}^{p_n} E(Y_i)^8 \leq K_2.$$

The assumption stating that the eighth moment of Y is bounded on the average is needed in the proof of the following fundamental result.

Theorem 7 *With notation as above*

$$lim_{n\to\infty}[E||\mathbf{S}_n - \mathbf{\Sigma}_n||_n^2 - \frac{p_n}{n}(m_n^2 + \theta_n^2)] = 0,$$

where

$$\theta_n^2 = Var\left(\frac{\sum_{i=1}^{p_n} Y_i^2}{p_n}\right),$$

is bounded in n provided that Assumption 2 is satisfied.

From this theorem it follows that in the following two important cases, the improved sample covariance matrix in (4.3) is a consistent estimator of $\mathbf{\Sigma}$.

Corollary 4.1 Under Assumptions 1–3, the estimator $\widehat{\mathbf{\Sigma}}^*$ is consistent if either

$$\frac{p_n}{n} \to 0,$$

or

$$m_n^2 + \theta_n^2 \to 0.$$

Note that the condition $m_n^2 \to 0$, in particular, implies that most of the p_n random variables must have asymptotically vanishing variances. This can be interpreted as a sort of *data sparsity*.

The last assumption involves the dependence structure of the data matrix through its PCs:

Assumption 3 Let Q_n denote the set of all quadruples that are made of four distinct integers between 1 and p_n, then

$$\lim_{n\to\infty}\frac{p_n^2}{n^2} \times \frac{\sum_{(i,j,k,l)\in Q_n}(\text{Cov}[Y_iY_j, Y_kY_l])^2}{\text{card}(Q_n)} = 0.$$

Details of proofs of the consistency results, further discussions on the assumptions and their geometric and Bayesian interpretations, and extensive simulation studies can be found in Ledoit and Wolf (2004). Their simulation results support the superiority of this class of estimators compared to the sample covariance matrix, the class of empirical Bayes estimators (Haff, 1980), and a few others reviewed in Lin and Perlman (1985).

There is a host of covariance estimators that are either inspired by or related to the shrinkage estimator given in (4.3). Some of which are reviewed briefly here.

- Fixing $\alpha_2 = 1$, Warton (2008) has shown that the resulting shrinkage/ridge estimators can be obtained using the penalized normal likelihood with a penalty term proportional to $\text{tr}\boldsymbol{\Sigma}^{-1}$ or the sum of inverse eigenvalues (see Chapter 5). Evidently, such a penalty forces the estimator to be a nonsingular matrix. The K-fold cross-validation of the likelihood function is used to select the penalty parameter. When the response variables are measured on different scales, it is more appropriate to regularize on the standardized scale. In such a situation, the sample correlation matrix $\widehat{R}$ is regularized as

$$\widehat{R}(\alpha) = \alpha\widehat{R} + (1-\alpha)I,$$

  where $\alpha \in [0, 1]$. For a related shrinkage estimator in the context of discriminant analysis, see Friedman (1989).
- A covariance estimator with an explicit constraint on its condition number (CN) is proposed by Won et al. (2009). Let $u_1, \cdots, u_p$ be the eigenvalues of $\boldsymbol{\Sigma}^{-1}$ and define $\widehat{\boldsymbol{\Sigma}}_{CN} = P\text{diag}(\widehat{u}_1^{-1}, \cdots, \widehat{u}_p^{-1})P'$ where P is the orthogonal matrix from the spectral decomposition of the sample covariance matrix $\mathbf{S} = P\text{diag}(\lambda_1, \cdots, \lambda_p)P'$. Then, the $\widehat{u}_1, \cdots, \widehat{u}_p$ are defined as the solution of the optimization problem:

$$\min_{u,u_1,\cdots,u_p} \sum_{i=1}^{p}(\lambda_i u_i - \log u_i) \quad \text{subject to} \quad -u \le u_i \le (\kappa_{\max})u, \tag{4.4}$$

  for $i = 1, \cdots, p$, where $\kappa_{\max}$ is a tuning parameter.
- A related alternative to the above proposal is to estimate a covariance matrix $\boldsymbol{\Sigma}$ through its matrix logarithm $A = \log\boldsymbol{\Sigma}$ (see Section 3.4), using a penalized normal log-likelihood function (Deng and Tsui, 2013). The matrix log-transformation provides the ability to impose a convex penalty on the modified likelihood such that the largest and smallest eigenvalues of the estimators can be regularized simultaneously. The penalty function is the Frobenius norm:

$$||A||_F^2 = \text{tr}(A^2) = \sum_{i=1}^{p}(\log\lambda_i)^2, \tag{4.5}$$

where λ_i is the ith eigenvalue of the covariance matrix $\boldsymbol{\Sigma}$. To appreciate the relevance of this penalty function, note that whether λ_i goes to zero or diverges to infinity, the value of $(\log \lambda_i)^2$ goes to infinity in both cases. Therefore, it simultaneously regularizes the largest and smallest eigenvalues of the covariance matrix estimate. This method transforms the problem of estimating the covariance matrix into the problem of estimating a symmetric matrix, which can be solved by an iterative quadratic programming algorithm.

- Applications of the shrinkage estimator in (4.3) to estimation of the spectral density matrix of multivariate stationary time series are discussed in Böhm and von Sachs (2008). Recently, the idea of a convex combination of a classical estimator with a target has been extended for estimating the spectral density matrix of a multivariate time series, which is the frequency-domain analog of the variance–covariance matrix. The shrinkage estimator for the spectral density matrix shrinks a nonparametric estimator to the scaled identity matrix so that the resulting estimate is well-conditioned. However, the off-diagonals of the estimator is shrunk to 0 and thus potentially biases the estimates of linear association toward the null. To overcome this problem, one may shrink a nonparametric estimator of the spectral density matrix toward a more general shrinkage target (Fiecas and Ombao, 2011).

4.2 REGULARIZING THE EIGENVECTORS

In this section, we present an overview of various methods for regularizing the eigenvectors or computing the sparse PCs. These usually lead to computationally challenging problems where the orthogonality of the regularized eigenvectors cannot be guaranteed.

In the high-dimensional data situations, the eigenvectors of the sample covariance matrix have too many entries, they are hard to interpret, and the traditional estimates of PCA loadings are inconsistent estimators of their population counterparts. Their use in practice could result in misleading conclusions (Johnstone and Lu, 2009). Therefore, regularization and imposing sparsity-inducing penalties on them could enhance their interpretability, statistical efficiency, and computability.

Recall that the standard PCA is based on *maximum variance property* of linear combinations of the random variables and involves computing the eigenvalues and eigenvectors of the sample covariance matrix $\mathbf{S}$ (see Section 3.5). Its goal is to find sequentially unit vectors $\mathbf{v}_1, \cdots, \mathbf{v}_p$ that maximize $\mathbf{v}'\boldsymbol{X}'\boldsymbol{X}\mathbf{v}$ subject to $\mathbf{v}_{i+1}$ being orthogonal to the previous vectors $\mathbf{v}_1, \cdots, \mathbf{v}_i$. Throughout, for simplicity the data matrix $\boldsymbol{X}$ is assumed to be column-centered so that the sample covariance matrix $\mathbf{S}$ is proportional to $\boldsymbol{X}'\boldsymbol{X}$.

Ideally, one would define a sparse vector of PC loadings $\mathbf{v}$ with a large number of zero entries as the solution of the following optimization problem:

$$\text{maximize}_{\mathbf{v}} \mathbf{v}'\boldsymbol{X}'\boldsymbol{X}\mathbf{v} \quad \text{subject to} \quad ||\mathbf{v}||_2^2 \leq 1, ||\mathbf{v}||_0 \leq k, \tag{4.6}$$

where the ℓ_0 norm $||\mathbf{v}||_0$ simply counts the number of nonzero entries of the vector $\mathbf{v}$ and k is an integer. The optimization amounts to seeking a linear combination of variables with the highest variance among those with at most k nonzero loadings. An advantage of imposing this penalty on $\mathbf{v}$ in (4.6) is that for a covariance matrix with sparse eigenvectors, its solutions are more sparse. This is similar to regression problems with many regressors, where imposing the Lasso penalty improves the performance of the standard least-squares regression.

The price to pay for this benefit is the computational cost, which is mostly due to the following basic facts:

1. The objective function $\mathbf{v}'X'X\mathbf{v}$ is convex in $\mathbf{v}$ and (4.6) calls for maximizing it. However, all tools in the convex optimization toolbox are designed for *minimizing* convex functions.
2. The ℓ_0 norm on $\mathbf{v}$ is discontinuous, nonconvex, and hard to work with.

Next, we present an overview of the research in recent years showing how the ℓ_0 norm can be relaxed and replaced by the ℓ_1 and other smooth norms.

The search for sparsity of standard PC loadings has led to a number of ad hoc thresholding methods. An approach proposed by Cadima and Jolliffe (1995) artificially sets standard PC loadings to zero if their absolute values are below a certain threshold. Another simplified component technique (SCoT) in which the PCA is followed by rotation was presented by Jolliffe and Uddin (2000). The SCoT procedure successively finds linear combinations of the variables that maximize a criterion which balances variance and a simplicity measure. Vines (2000) has considered PCs with loadings restricted to only integers such as 0, 1, and -1.

A more principled group of procedures, referred to collectively as sparse PCA methods, aims at finding loading vectors with many zero entries. Thus, increasing the interpretability of PCs by reducing the number of explicitly used variables. Conceptually, the simplest and most natural example of these due Jolliffe et al. (2003) is the SCoTLASS procedure. It is the first attempt to replace the ℓ_0 norm in (4.6) by the Lasso penalty on the PC loadings. More precisely, SCoTLASS finds the first vector of PC loadings by solving the optimization problem:

$$\text{maximize}_{\mathbf{v}} \mathbf{v}X'X\mathbf{v} \quad \text{subject to} \quad ||\mathbf{v}||_2^2 \leq 1, \quad ||\mathbf{v}||_1 \leq c. \tag{4.7}$$

The subsequent PCs solve the same problem, but with the additional constraint that $\mathbf{v}_{i+1}$ must be *orthogonal* to the previous PCs $\mathbf{v}_1, \cdots, \mathbf{v}_i$.

Computationally, SCoTLASS is a difficult problem since it still requires *maximizing* a convex objective function instead of the traditional task of *minimizing* it. However, it can be shown that (Witten et al., 2009) the solution of (4.7) can be obtained from solving the following problem involving the data matrix itself:

$$\text{maximize}_{\boldsymbol{u},\mathbf{v}} \boldsymbol{u}'X\mathbf{v} \quad \text{subject to} \quad ||\mathbf{v}||_1 \leq c, \quad ||\mathbf{v}||_2^2 \leq 1, \quad ||\boldsymbol{u}||_2^2 \leq 1. \tag{4.8}$$

We refer to this problem as the *sparse principal component* or SPC in the sequel. The new optimization problem is biconvex in $\boldsymbol{u}$ and $\mathbf{v}$, in the sense that for a fixed $\mathbf{v}$ it is convex in $\boldsymbol{u}$ and vice versa. To see the connection between the two optimization problems, for the moment fix $\mathbf{v}$ and consider maximizing the SPC criterion (4.8) over $\boldsymbol{u}$. Applying the Cauchy–Schwartz inequality to $\boldsymbol{u}'(\boldsymbol{X}\mathbf{v})$ in (4.8), and recalling that the equality is attained for a $\boldsymbol{u}$ given by $\frac{\boldsymbol{X}\mathbf{v}}{||\boldsymbol{X}\mathbf{v}||_2}$, it follows that a $\mathbf{v}$ that maximizes (4.8) also maximizes

$$\mathbf{v}'\boldsymbol{X}'\boldsymbol{X}\mathbf{v} \quad \text{subject to} \quad ||\mathbf{v}||_1 \leq c, \quad ||\mathbf{v}||_2 \leq 1. \tag{4.9}$$

Of course, this can be recognized as the SCoTLASS criterion (4.7), and the SPC Algorithm given below (Witten et al., 2009) can be used to find the SCoTLASS solution, but, without enforcing the orthogonality of the regularized eigenvectors.

The SPC Algorithm (Witten et al., 2009):

---

1. Initialize $\mathbf{v}$ to have ℓ_2 norm 1.
2. Iterate until convergence:
   (a) $\boldsymbol{u} \leftarrow \frac{\boldsymbol{X}\mathbf{v}}{||\boldsymbol{X}\mathbf{v}||_2}$,
   (b) $\mathbf{v} \leftarrow \frac{S(\boldsymbol{X}'\boldsymbol{u},\Delta)}{||S(\boldsymbol{X}'\boldsymbol{u},\Delta)||_2}$, where $S(\cdot,\cdot)$ is the soft-thresholding operator, and $\Delta = 0$ if the computed $\mathbf{v}$ satisfies $|||\mathbf{v}||_1 \leq c$; otherwise, a positive constant Δ is chosen such that $||\mathbf{v}||_1 = c$.

The **R** package **PMA** (penalized multivariate analysis) developed by Witten et al. (2009) implements the above algorithm and a few other related procedures in multivariate analysis.

4.3 A DUALITY BETWEEN PCA AND SVD

In this section, we present the SVD of a rectangular matrix and discuss the role of its low-rank approximation property in revealing a duality between the PCA and SVD. This duality is central in converting the more challenging computational problem (4.6) of maximizing a convex function to that of minimizing a norm or developing a regression-based approach to sparse PCA and SVD of a data matrix.

Theorem 8 (The Singular Value Decomposition) *Let $\boldsymbol{X}$ be an $n \times p$ matrix of rank r. Then,*

(a) there exist matrices $\mathbf{U}$, $\mathbf{V}$, *and* $\mathbf{D}$, *such that*

$$\boldsymbol{X} = \mathbf{UDV}' = \sum_{k=1}^{r} d_k \boldsymbol{u}_k \mathbf{v}_k', \tag{4.10}$$

where the columns of $\mathbf{U} = (\boldsymbol{u}_1, \cdots, \boldsymbol{u}_r)$, $\mathbf{V} = (\mathbf{v}_1, \cdots, \mathbf{v}_r)$ *are orthonormal, and the diagonal entries of* $\mathbf{D} = diag(d_1, \cdots, d_r)$ *are ordered nonnegative numbers, that is,* $d_1 \geq d_2 \cdots \geq d_r > 0$.

(b) (Eckart–Young Theorem) for any integer $l \leq r$, *the matrix*

$$\boldsymbol{X}^{(l)} = \sum_{k=1}^{l} d_k \boldsymbol{u}_k \mathbf{v}_k', \tag{4.11}$$

is the closest rank-l approximation to $\boldsymbol{X}$ *in the Frobenius norm, that is*

$$\boldsymbol{X}^{(l)} = \underset{\mathbf{X}^* \in \mathcal{M}_{(l)}}{\text{argmin}} \, ||\boldsymbol{X} - \boldsymbol{X}^*||_F^2,$$

where $\mathcal{M}_{(l)}$ *is the set of all* $n \times p$ *matrices of rank* l.

The columns of $\mathbf{U}$ and $\mathbf{V}$ are called the *left and right singular vectors* of $\boldsymbol{X}$, respectively, and the diagonal entries of $\mathbf{D}$ are the corresponding *singular values* (Golub and Van Loan, 1996). The SVD (4.10) represents $\boldsymbol{X}$ as the sum of r orthogonal *layers* of decreasing importance. For the kth layer $d_k \boldsymbol{u}_k \mathbf{v}_k'$ the entries of $\boldsymbol{u}_k$ and $\mathbf{v}_k$ represent the subjects and responses (variables) effects, respectively. It is common to focus on the SVD layers corresponding to larger d_k values and ignore the rest or treat them as the noise.

The following identities are easy to verify using (4.10) and the orthogonality of the singular vectors:

$$\boldsymbol{X}'\boldsymbol{X} = \mathbf{VD}^2\mathbf{V}', \quad \boldsymbol{X}\mathbf{V} = \mathbf{UD} = (\boldsymbol{X}\mathbf{v}_1, \cdots, \boldsymbol{X}\mathbf{v}_r). \tag{4.12}$$

The first identity reveals that the columns of $\mathbf{V}$ are, indeed, the vectors of the PC loadings and the second shows that the columns of $\mathbf{UD}$ are the corresponding PCs of the data matrix. Thus, imposing Lasso penalty on the columns of $\mathbf{V}$ as in Zou et al. (2006) will help produce sparse PCA. Interestingly, the three components $(\mathbf{U}, \mathbf{V}, \mathbf{D})$ of the SVD of $\boldsymbol{X}$ can be obtained from the spectral decompositions of the two matrices $\boldsymbol{X}\boldsymbol{X}'$ and $\boldsymbol{X}'\boldsymbol{X}$:

$$\boldsymbol{X}\boldsymbol{X}' = \mathbf{UD}^2\mathbf{U}' \quad \text{and} \quad \boldsymbol{X}'\boldsymbol{X} = \mathbf{VD}^2\mathbf{V}'. \tag{4.13}$$

It should be noted that compared to PCA, SVD is a more fundamental tool in the sense that it simultaneously provides the PCs for both the row and column spaces.

The low-rank approximation property of the SVD is the key to developing a regression-based approach to sparse PCA (Shen and Huang, 2008) as the following result shows.

Lemma 7 *Let* $\boldsymbol{u}$ *and* $\mathbf{v}$ *be* n*- and* p*-vectors with* $||\boldsymbol{u}|| = 1$ *and* d *be a nonnegative number. Then,*

(a)

$$||\boldsymbol{X} - d\boldsymbol{u}\mathbf{v}'||_F^2 = ||\boldsymbol{X}||_F^2 - 2d\boldsymbol{u}'\boldsymbol{X}\mathbf{v} + d^2. \tag{4.14}$$

(b) approximating the data matrix $\boldsymbol{X}$ *by rank-one matrices of the form* $d\boldsymbol{u}\mathbf{v}'$ *in the Frobenius norm or minimizing the penalized sum of squares criterion*

$$||\boldsymbol{X} - d\boldsymbol{u}\mathbf{v}'||_F^2 + P_\lambda(\boldsymbol{u}, \mathbf{v}), \tag{4.15}$$

with respect to the triplet $(d, \boldsymbol{u}, \mathbf{v})$ *is equivalent to maximizing the penalized bilinear form*

$$\boldsymbol{u}'\boldsymbol{X}\mathbf{v} - P_\lambda(\boldsymbol{u}, \mathbf{v}), \tag{4.16}$$

where $P_\lambda(\cdot, \cdot)$ *is a penalty function with* λ *its tuning parameter.*

The proof of (a) is simple and follows from expanding the left-hand side of the identity using the definition of the Frobenius norm. From the identity (4.14), it is evident that minimizing its left-hand side over $(d, \boldsymbol{u}, \mathbf{v})$ under some constraints is equivalent to maximizing the bilinear form $\boldsymbol{u}'\boldsymbol{X}\mathbf{v}$ under the same constraints (see Section 1.4). This observation is central to solving a variety of sparse PCA, SVD, penalized matrix decomposition (PMD), and regression problems in a unified manner using a variety of penalty functions listed below.

- Additive penalty (Shen and Huang, 2008):

$$P_\lambda(\boldsymbol{u}, \mathbf{v}) = d(\lambda_u||\boldsymbol{u}||_1 + \lambda_v||\mathbf{v}||_1), \tag{4.17}$$

  where λ_u and λ_v are the tuning parameters for the left and right singular vectors, respectively. Using two penalty parameters allows different levels of sparsity on the two singular vectors. Setting $\lambda_u = 0$ will lead to sparse PCA.
- Multiplicative penalty (Chen et al., 2012):

$$P_\lambda(\boldsymbol{u}, \mathbf{v}) = \lambda \sum_{i=1}^{n}\sum_{j=1}^{p} w_{ij}|du_iv_j| = \lambda||\boldsymbol{u}||_{1,w^{(u)}}||\mathbf{v}||_{1,w^{(v)}} \tag{4.18}$$

  where $w_{ij} = w^{(d)}w_i^{(u)}w_j^{(v)}$ are certain data-driven weights. A more general form of the multiplicative penalty function is used by Huang et al. (2009) in the context of two-way functional data.

- PMD-SVD (Witten et al., 2009):

$$||\boldsymbol{u}|| \leq 1, ||\mathbf{v}|| \leq 1, P_1(\boldsymbol{u}) \leq c_1, P_2(\mathbf{v}) \leq c_2, \tag{4.19}$$

where $P_i(\cdot)$, $i = 1, 2$, are possibly different penalty functions such as Lasso and fused Lasso, and c_i, $i = 1, 2$, are the tuning parameters.

The more general problem of approximating the data matrix $\boldsymbol{X}$ by the best rank-l matrices is done sequentially by first finding the best rank-one matrix of the form $d\boldsymbol{u}\mathbf{v}'$. Then, the subsequent triplets $\{d_k, \boldsymbol{u}_k, \mathbf{v}_k\}$, $k > 1$, are found as the rank-one approximation of the corresponding residual matrices, that is, $d_2\boldsymbol{u}_2\mathbf{v}_2'$ is the best rank-one approximation of $\boldsymbol{X} - d_1\boldsymbol{u}_1\mathbf{v}_1'$, and so on. Because of the penalty imposed on them, the computed singular vectors are no longer orthogonal as in the classical SVD.

The singular values are usually assumed to be distinct so that the SVD decomposition is unique up to the signs of the singular vectors. If some singular values are identical, then the preceding SVD decomposition is not unique, which could complicate the theoretical analysis of the procedure. However, the recent orthogonal iteration approach for estimating the principal subspaces spanned by the first few eigen- and singular-vectors avoids the computational difficulties and guarantees the orthogonality of the singular vectors (Ma, 2011; Yang et al., 2011).

4.4 IMPLEMENTING SPARSE PCA: A DATA EXAMPLE

In this section, we discuss a few consequences of the lack of orthogonality of the sparse eigenvectors. When the right singular vector of $\boldsymbol{X}$ is known to be sparse, it is prudent to penalize it using the Lasso penalty, $P_\lambda(\boldsymbol{u}, \mathbf{v}) = \lambda d||\mathbf{v}||_1$. A key difference between the standard and sparse PCA is that, in the latter the **PC** loading vectors $\mathbf{v}_i$'s are not orthogonal and hence the PCs $\mathbf{Z}_i = \boldsymbol{X}\mathbf{v}_i$ are correlated. Thus, special attention should be paid to selecting the leading PCs based on the usual proportion of variance explained.

In the standard PCA, the ith PC $\mathbf{Z}_i = \boldsymbol{X}\mathbf{v}_i$ is obtained as the projection of the data matrix onto the loading vector $\mathbf{v}_i$. The variance explained by the ith PC $\mathbf{Z}_i$ is $||\mathbf{Z}_i||^2$, and the variance explained by the first k PCs is defined as $\text{tr}\mathbf{Z}_{(k)}'\mathbf{Z}_{(k)} = \sum_{i=1}^{k} ||\mathbf{Z}_i||^2$ which is additive. In sparse PCA, however, we need to adjust the proportion of variance explained to reflect its loss of orthogonality. For nonorthogonal loading vectors it is not advisable to compute separate projection of the data onto each of the first k loading vectors (Shen and Huang, 2008). Instead, one may project the data matrix $\boldsymbol{X}$ onto the k-dimensional subspace generated by the column space of $\mathbf{V}_k = (\mathbf{v}_1, \cdots, \mathbf{v}_k)$, that is,

$$\boldsymbol{X}_k = \boldsymbol{X}\mathbf{V}_k(\mathbf{V}_k'\mathbf{V}_k)^{-1}\mathbf{V}_k'.$$

Then, the total variance explained by the first k sparse PCs is defined as $\text{tr}(\boldsymbol{X}_k'\boldsymbol{X}_k)$. It is easy to verify that when the $\mathbf{v}_i$'s are orthonormal, then these modified definitions reduce to their standard counterparts.

It can be shown that the new concept of explained variance is monotone in k, that is

$$\operatorname{tr}(\boldsymbol{X}_k'\boldsymbol{X}_k) \leq \operatorname{tr}(\boldsymbol{X}_{k+1}'\boldsymbol{X}_{k+1}) \leq \operatorname{tr}(\boldsymbol{X}'\boldsymbol{X}). \tag{4.20}$$

Thus, we define the *cumulative percentage of explained variance* (CPEV) as

$$\operatorname{tr}(\boldsymbol{X}_k'\boldsymbol{X}_k)/\operatorname{tr}(\boldsymbol{X}'\boldsymbol{X}),$$

which is always in the interval [0, 1]. The plot of CPEV versus k, the number of sparse PCs, can be used as a modified *scree plot* to choose the number of important PCs.

The tuning parameter $\lambda = \lambda_v$ can be selected via an approach similar to the cross-validation as follows:

1. Randomly split the rows of the data into m groups of roughly equal size to obtain m data matrices $\boldsymbol{X}^1, \cdots, \boldsymbol{X}^m$.
2. For each candidate value of λ_v from a finite list,
   (a) and each $\boldsymbol{X}^i$, $1 \leq i \leq m$, fit the SPC to $\boldsymbol{X}^i$ with tuning parameter λ_v, and calculate the estimate of $\boldsymbol{X}^i$ as $\widehat{\boldsymbol{X}}^i = d\boldsymbol{u}\boldsymbol{v}'$. Then compute the mean squared error of the estimate $\widehat{\boldsymbol{X}}^i$ which is the mean of the squared differences between elements of $\boldsymbol{X}^i$ and the corresponding elements of $\widehat{\boldsymbol{X}}^i$, where the mean is taken only over elements that are missing from $\boldsymbol{X}^i$.
   (b) Compute the average mean squared error across $\boldsymbol{X}^1, \cdots, \boldsymbol{X}^m$ for tuning parameters λ_v.
3. Choose the value(s) of λ_v with the smallest average mean squared error.

We illustrate the methodology using a classic dataset and the package **PMA** in **R**.

Example 30 (Pitprops Data) *The Pitprops dataset, first analyzed by Jeffers (1967), has become a standard benchmark and a classic example of the difficulty of interpreting fully loaded factors with standard PCA. This dataset is concerned with the maximum compressive strength of the props and involves 13 physical variables that can be measured on the pitprops. The measurements were made on a sample of* 180 *pitprops and only the correlation matrix of the data is available. The PCA is conducted on the correlation matrix. Table 4.1 gives the loadings and the proportions of explained variance by the first six PCs. The cumulative percentage of explained variance (CPEV) by these PCs is 87%.*

We also apply the SPC algorithm (Witten et al., 2009) to the Pitprops data. We select the tuning parameter minimizing the mean squared error via 10-fold cross-validation and obtain the optimal value $\widehat{\lambda_v} = 3.312$. *Table 4.2 contains the first six sparse SPCs with* $\lambda_v = 3.312$ *and the proportions of variance explained by the first k SPCs. Compared with the standard PCA, the sparse PCs only have three 0 elements and the CPEV for the first six PCs is comparable to the one for the standard PCA (83.6 vs. 87).*

TABLE 4.1 Loadings of the first six PCs by PCA

	PC1	*PC2*	*PC3*	*PC4*	*PC5*	*PC6*
topdiam	−0.404	−0.218	0.207	0.091	−0.083	0.120
length	−0.406	−0.186	0.235	0.103	−0.113	0.163
moist	−0.124	−0.541	−0.141	−0.078	0.350	−0.276
testsg	−0.173	−0.456	−0.352	−0.055	0.356	−0.054
ovensg	−0.057	0.170	−0.481	−0.049	0.176	0.626
ringtop	−0.284	0.014	−0.475	0.063	−0.316	0.052
ringbut	−0.400	0.19	−0.253	0.065	−0.215	0.003
bowmax	−0.294	0.189	0.243	−0.286	0.185	−0.055
bowdist	−0.357	−0.017	0.208	−0.097	−0.106	0.034
whorls	−0.379	0.248	0.119	0.205	0.156	−0.173
clear	0.011	−0.205	0.070	−0.804	−0.343	0.175
knots	0.115	−0.343	−0.092	0.301	−0.600	−0.170
diaknot	0.113	−0.309	0.326	0.303	0.080	0.626
CPEV	32.5	50.7	65.2	73.7	80.7	87

4.5 SPARSE SINGULAR VALUE DECOMPOSITION (SSVD)

In this section, we go one step beyond the sparse PCA by penalizing both the left and right singular vectors of the SVD of a data matrix $\boldsymbol{X}$ and discuss its application to the emerging topic of biclustering.

The goal of sparse SVD is to find a low-rank matrix approximation to the data matrix $\boldsymbol{X}$ of the form of (4.10), but requiring that the vectors $\boldsymbol{u}_i$ and $\mathbf{v}_i$ to be sparse or have many zero entries. This can be achieved using a suitable sparsity-inducing penalty in expression (4.15). Then, the sparse rank-one matrix $d_k \boldsymbol{u}_k \mathbf{v}_k$ is called the

TABLE 4.2 Loadings of the first six sparse PCs by the SPC

	PC1	*PC2*	*PC3*	*PC4*	*PC5*	*PC6*
topdiam	−0.308	−0.391	0.049	0.099	−0.006	0.076
length	−0.319	−0.379	0	0.118	−0.016	0.089
moist	0.068	−0.358	0.478	−0.146	−0.214	−0.225
testsg	0	−0.166	0.587	−0.106	−0.094	−0.297
ovensg	0	0.545	0.114	0.009	−0.373	−0.39
ringtop	−0.206	0.271	0.376	0.089	−0.17	0.315
ringbut	−0.409	0.227	0.138	0.094	−0.192	0.235
bowmax	−0.289	−0.008	−0.334	−0.254	−0.295	−0.149
bowdist	−0.306	−0.192	−0.127	−0.07	−0.144	0.105
whorls	−0.423	0.042	−0.18	0.215	−0.364	−0.082
clear	0.16	−0.072	−0.101	−0.764	−0.366	0.282
knots	0.333	−0.065	0.114	0.331	−0.404	0.562
diaknot	0.309	−0.278	−0.262	0.343	−0.45	−0.325
CPEV	29.1	45.6	60.6	69.1	76.7	83.6

kth SSVD *layer* where the nonzero entries u_{ik} of the kth left singular vector and the nonzero entries v_{ik} of the right singular vector could introduce a checkerboard pattern useful for biclustering. Therefore, the kth layer can simultaneously link sets of subjects (samples) and sets of variables and reveal interesting sample–variable (row–column) association.

A penalized least-squares criterion for obtaining SSVD layers proposed by Lee et al. (2010) is based on the adaptive Lasso penalties (Zou, 2006). We use the simpler Lasso penalties here and obtain the first pair of sparse singular vectors by minimizing

$$||\boldsymbol{X} - d\boldsymbol{u}\mathbf{v}'||_F^2 + d(\lambda_u||\boldsymbol{u}||_1 + \lambda_v||\mathbf{v}||_1), \text{ subject to } \quad ||\boldsymbol{u}||_2 = ||\mathbf{v}||_2 = 1, \quad (4.21)$$

with respect to $(d, \boldsymbol{u}, \mathbf{v})$. Note that for $\boldsymbol{u}$ fixed with $||\boldsymbol{u}|| = 1$, the optimization problem (4.21) is the same as that in (1.14). Thus, a solution can be obtained by alternating between the following two steps till convergence:

$$\begin{aligned} (1) \quad \mathbf{v} &= \frac{S(\boldsymbol{X}'\boldsymbol{u}, \lambda_u)}{||S(\boldsymbol{X}'\boldsymbol{u}, \lambda_u)||_2}, \\ (2) \quad \boldsymbol{u} &= \frac{S(\boldsymbol{X}'\mathbf{v}, \lambda_v)}{||S(\boldsymbol{X}'\mathbf{v}, \lambda_v||_2}, \end{aligned}$$

and at the convergence we set $d = \boldsymbol{u}'\boldsymbol{X}\mathbf{v}$. Once the triplet $(d_1, \boldsymbol{u}_1, \mathbf{v}_1)$ for the first rank-one approximation is found, the others are obtained sequentially by applying the procedure to the residual matrix $\boldsymbol{X} - d_1\boldsymbol{u}_1\mathbf{v}_1'$, and so on (see Lee et al., 2010).

A slightly different optimization problem for rank-one approximation considered by Witten et al. (2009) is to minimize

$$||\boldsymbol{X} - d\boldsymbol{u}\mathbf{v}'||_F^2, \text{ subject to } \quad ||\boldsymbol{u}||_2 = ||\mathbf{v}||_2 = 1, ||\boldsymbol{u}||_1 \le c_u, ||\mathbf{v}||_1 \le c_v. \quad (4.22)$$

The solution is obtained by iterating between the following two steps:

$$\begin{aligned} (1) \quad \mathbf{v} &= \frac{S(\boldsymbol{X}'\boldsymbol{u}, \Delta_u)}{||S(\boldsymbol{X}'\boldsymbol{u}, \Delta_u)||_2}, \\ (2) \quad \boldsymbol{u} &= \frac{S(\boldsymbol{X}'\mathbf{v}, \Delta_v)}{||S(\boldsymbol{X}'\mathbf{v}, \Delta_v||_2}, \end{aligned}$$

where Δ_u, Δ_v are chosen by a search algorithm such that $||\boldsymbol{u}||_1 = c_u, ||\mathbf{v}||_1 = c_v$, and at the convergence we set $d = \boldsymbol{u}'\boldsymbol{X}\mathbf{v}$ (see the SPC Algorithm).

For a given data matrix $\boldsymbol{X}$ of high dimension, one of the motivations for inducing sparsity on the singular vectors comes from *biclustering*, where the goal is to find certain "checkerboard" patterns in the data matrices or to identify sets of rows and columns that are significantly associated (Lee et al., 2010). In a genomic study, the rows of $\boldsymbol{X}$ could be subjects who might be healthy or have a number of types of cancers, and its columns correspond to a large number of genes. In such studies, the goal is to simultaneously identify genes and subject groups that are related. In other words, one is looking for groups of genes that are significantly expressed for certain types of cancer, or that can help distinguish different types of cancer.

The lung cancer data (Lee et al., 2010) consist of expression levels of $p = 12,625$ genes, measured from $n = 56$ subjects where the subjects are known to be either normal subjects (Normal) or patients with one of the following three types of cancer: pulmonary carcinoid tumors (Carcinoid), colon metastases (Colon), and small cell carcinoma (SmallCell). The data matrix $\boldsymbol{X}$ here is $56 \times 12,625$ whose rows represent the subjects, grouped together according to the cancer type, and the columns correspond to the genes. It is expected that a subset of the genes would regulate the types of cancers, therefore the singular vectors corresponding to the genes should ideally be sparse. Each column of $\boldsymbol{X}$ is centered before the SSVD analysis. Lee et al. (2010) extract sequentially the first three SSVD layers $\boldsymbol{X}_{(k)} = d_k \boldsymbol{u}_k \mathbf{v}_k'$, since the first three singular values are much bigger than the rest. The degree of sparsity or the number of zero entries in $\mathbf{v}_k$'s turns out to be disappointingly small for the methods proposed by Lee et al. (2010), Witten et al. (2009), and Yang et al. (2011). A slightly better sparsity result for these data is obtained by Chen et al. (2012) using penalized multivariate regression and exploiting the cancer type as a covariate.

4.6 CONSISTENCY OF PCA

In the high-dimensional data situations, the standard PCA and SVD have rather poor statistical properties. The high level of noise in such situations can overwhelm the signal to the extent that the standard estimates of PCA and SVD loadings are not consistent for their population counterparts. Thus, it is natural to study the consistency of PCA under reasonable sparsity assumptions on the population eigenvectors and the covariance matrix. In this section, models, conditions, and results for consistency of PCA in high-dimensional setup are presented following the lead of Johnstone and Lu (2009).

Consider a single factor or spiked covariance model

$$\boldsymbol{x}_i = f_i \mathbf{v} + \sigma \mathbf{z}_i, i = 1, \cdots, n, \tag{4.23}$$

where $\mathbf{v} \in R^p$ is the eigenvector or single component to be estimated, $f_i \sim N(0, 1)$ are i.i.d. normal random effects, and $\mathbf{z}_i \sim N_p(0, I)$ are independent noise vectors. Note that the notation in the factor model here is slightly different from that in (1.21). Let $\widehat{\mathbf{v}}$ be the eigenvector of $\mathbf{S}$ associated with the largest sample eigenvalue. A natural measure of closeness of $\widehat{\mathbf{v}}$ to $\mathbf{v}$ is the "overlap" or the cosine of the angle between them:

$$R(\widehat{\mathbf{v}}, \mathbf{v}) = \frac{\widehat{\mathbf{v}}' \mathbf{v}}{||\widehat{\mathbf{v}}||_2 ||\mathbf{v}||_2}.$$

For the asymptotic analysis, it is common to assume that there is a sequence of models (4.23) indexed by n where the model parameters $p_n, \mathbf{v}_n, \sigma_n$ are allowed to depend on n, though for simplicity this is not shown explicitly in some formulae. The vector of sample PC loadings $\widehat{\mathbf{v}}_n$ is said to be consistent for $\mathbf{v}$, its population counterpart, provided that

$$R(\widehat{\mathbf{v}}_n, \mathbf{v}_n) \to 1, \quad \text{as} \quad n \to \infty.$$

It turns out that the consistency of PCA depends critically on the limiting values of

$$\lim p_n/n = c, \quad \text{and} \quad \lim ||\mathbf{v}_n||_2^2/\sigma^2 = \omega > 0, \tag{4.24}$$

where ω stands for the limiting value of the *signal-to-noise ratio* in the model.

Theorem 9 *Assume that (4.24) holds for n observations drawn from the p-dimensional model (4.23). Then,*

(a) almost surely

$$\lim R^2(\widehat{\mathbf{v}}, \mathbf{v}) = R_\infty^2(\omega, c) = \frac{(\omega^2 - c)_+}{\omega^2 + c\omega}. \tag{4.25}$$

(b) $R_\infty^2(\omega, c) < 1$, *if and only if* $c > 0$.
(c) $R_\infty^2(\omega, c) = 1$ *and the sample PC* $\widehat{\mathbf{v}}$ *is a consistent estimator of* $\mathbf{v}$, *if and only if* $c = 0$.
(d) $R_\infty^2(\omega, c) = 0$ *if* $\omega^2 \le c$.
In this case, $\widehat{\mathbf{v}}$ *and* $\mathbf{v}$ *are asymptotically orthogonal so that* $\widehat{\mathbf{v}}$ *eventually carries no information about* $\mathbf{v}$.

The message is that when $p \gg n$ or when there are too many variables relative to the sample size n, the noise does not average out and the sample eigenvector is not consistent for its population counterpart. This suggests that an initial screening and selection of a relatively small subset of the original p variables might reduce the noise level and be beneficial before any PCA is attempted. How does one perform this needed screening of the variables? Intuitively, the screening can be more successful if the population PC $\mathbf{v}$ is sparse or concentrated with a large number of zero entries in the sense made precise next.

Assume that the rows of the data matrix $\boldsymbol{X}$, and the population PC can be represented in terms of a fixed orthonormal basis $\{\mathbf{e}_\nu\}$:

$$x_i = \sum_{\nu=1}^{p} x_{i,\nu}\mathbf{e}_\nu, \quad i = 1, \cdots, n, \quad \mathbf{v} = \sum_{\nu=1}^{p} v_\nu \mathbf{e}_\nu,$$

where the index ν will indicate the transform domain. The choice of the orthonormal basis and transform domain will depend on the dataset and the application area.

In cases of time series, longitudinal, and functional data, the x_i's are collected in the time domain, then the finite Fourier transform and wavelet transform and their corresponding basis have proved useful for various applications.

The idea of sparsity of $\mathbf{v}$ can be quantified by considering its ordered coefficient magnitudes

$$|v|_{(1)} \geq |v|_{(2)} \geq \cdots$$

and finding a relatively small integer k so that the sum of squares of the first largest k coordinates $\sum_{i=1}^{k} v_{(i)}^2$ is close to $||\mathbf{v}||^2 = \sum_{i=1}^{p} v_i^2$. This can only happen when the magnitudes of the coefficients decay quickly like

$$|v|_{(\nu)} \leq C\nu^{-1/q}, \ \nu = 1, 2, \ldots \tag{4.26}$$

for some $0 < q < 2$ and $C > 0$. Note that the smaller values of q force the coefficients to decay more quickly.

Next, it is shown that if the PCs have a sparse representation in terms of a basis $\{\mathbf{e}_\nu\}$, then selection of a suitable subset of variables should overcome the inconsistency of the standard PCA described in Theorem 9.

In model (4.23), assuming that σ^2 is known, the sample variances

$$\widehat{\sigma}_\nu^2 = n^{-1} \sum_{i=1}^{n} x_{i\nu}^2 \sim (\sigma^2 + v_\nu^2)\chi_n^2/n,$$

so that components ν with large values of v_ν are expected to have large sample variances. Thus, a suitable selection rule for a subset of the p variables is

$$\widehat{I} = \{\nu : \widehat{\sigma}_\nu^2 \geq \sigma^2(1 + \alpha_n)\} \tag{4.27}$$

where $\alpha_n = \alpha(n^{-1} \log(n \vee p))^{1/2}$ and α is a sufficiently large positive constant. Once $\widehat{I}$ is chosen, apply the standard PCA to $\mathbf{S}_I$ the submatrix of the sample covariance matrix $\mathbf{S}$, giving the leading eigenvector $(v_\nu, \nu \in \widehat{I})$. The corresponding sparse eigenvector of $\mathbf{S}$, denoted by $\mathbf{v}_I$, has entry v_ν if $\nu \in \widehat{I}$, and zero otherwise.

Theorem 10 *Consider the single factor model (4.23) with* $\log(p \vee n)/n \to 0$, $||\mathbf{v}_n|| \to ||\mathbf{v}||$ *as* $n \to \infty$ *and condition (4.26) satisfied. Then, the estimated leading eigenvector* $\mathbf{v}_I$ *obtained via the subsect selection rule (4.27) is consistent, that is,*

$$R(\widehat{\mathbf{v}}_I, \mathbf{v}) \to 1, \quad \textit{almost surely}.$$

The theorem shows the critical importance of the existence of a sparse representation for $\mathbf{v}$, so that it can be approximated (consistently) using a small number of coefficients via a PCA of a reduced data matrix. The ideas and techniques leading to Theorem 10 are summarized in the following generic algorithm, referred to as adaptive sparse principal component analysis (ASPCA). It does not require the specification of any particular model such as the single factor model (4.23) or more general factor models.

The ASPCA Algorithm:

---

1. *Compute Basis Coefficients*: Given a basis $\{\mathbf{e}_\nu\}$ for R^p, compute coordinates $x_{i\nu} = (\boldsymbol{x}_i, \mathbf{e}_\nu)$ in this basis for each $\boldsymbol{x}_i, i = 1, \cdots, n$.
2. *Subset*: Compute the sample variances $\widehat{\sigma}_\nu^2$ for the coordinates $\{x_{i\nu}\}$ for each ν. Let $\widehat{I}$ denote the set of indices ν corresponding to the largest k variances $\widehat{\sigma}_\nu^2$.
3. *Reduced PCA*: Apply the standard PCA to the reduced dataset $\{x_{i\nu} : \nu \in \widehat{I}, i = 1, \cdots, n\}$ to obtain the eigenvectors $\widehat{\mathbf{v}}_j = (\widehat{v}_{\nu,j}), j = 1, \cdots, k, \nu \in \widehat{I}$.
4. *Thresholding*: Filter out the noise in the estimated eigenvectors by soft-thresholding:

$$\tilde{v}_{\nu,j} = S(\widehat{v}_{\nu,j}, \lambda_j).$$

5. *Reconstruction*: Return to the original data domain by backtransforming $\tilde{\mathbf{v}}_\nu, \nu \in \widehat{I}$ to $\widehat{\mathbf{v}}_j$, using the given basis $\{\mathbf{e}_\nu\}$.

The need for the thresholding step in the ASPCA algorithm was not discussed earlier in this section. Through several examples it was found by Johnstone and Lu (2009) that thresholding successfully filters the noise in the data. The more prominent role of thresholding has been pointed out by Yang et al. (2011). There are a number of ways to choose the thresholds λ_j. The more formal choice suggested in analogy with the Gaussian sequence model is $\lambda_j = \widehat{\tau}_j\sqrt{2\log k}$, where $\widehat{\tau}_j$ is an estimate of the noise level in $\{\widehat{v}_{\nu,j}, \nu \in \widehat{I}\}$ given by

$$\widehat{\tau}_j = \text{MAD}\{\widehat{v}_{\nu,j}, \nu \in \widehat{I}\}/0.6745,$$

where MAD stands for the mean absolute deviation.

The key assumption in the ASPCA approach of Johnstone and Lu (2009) is that the individual signals as well as the eigenvectors are simultaneously sparse in some unknown basis. An alternative to ASPCA is the thresholded covariance estimator (Bickel and Levina, 2008b), the key assumption is the sparsity of the covariance matrix itself and not its eigenvectors. Consistency of the thresholded covariance estimator in the operator (spectral) norm implies the consistency of their eigenvectors and the PCA (see Theorem 5 and Bickel and Levina, 2008b). A proposal by Nadler (2009) which is in the spirit of the ASPCA Algorithm relies on the structure of the sample correlation matrix.

4.7 PRINCIPAL SUBSPACE ESTIMATION

In this section, Steps 2–4 of the ASPCA algorithm combined with orthogonal iterations will be used to develop a fast algorithm for computing the subspace spanned by the first few leading singular vectors. The end result is the *sparsely initialized thresholded subspace iterations* algorithm for sparse PCA and SVD of a large data matrix. It is computationally faster and performs statistically better than the existing SVD methods in the literature.

The algorithm makes effective use of the sparsity at the initialization stage using the subset selection idea in Johnstone and Lu (2009). It even beats the standard SVD in terms of computational speed since it estimates the threshold parameters by employing asymptotic results from the Gaussian sequence models and hence avoids the computationally expensive cross-validation technique. Finally, the orthogonality of the estimated eigenvectors is ensured using iterated projections described next.

For a $p \times p$ positive-definite matrix A, the *orthogonal iterations* (Golub and Van Loan, 1996) is a standard method for computing its leading r-dimensional eigenspace. For $r = 1$, the method reduces to the familiar *power method*. It starts with a $p \times r$ orthonormal matrix $Q^{(0)}$ and generates a sequence of $p \times r$ orthonormal matrices $Q^{(k)}, k = 1, 2, \cdots$, by alternating between the following two steps until convergence:

1. Multiplication: $T^{(k)} = AQ^{(k-1)}$,
2. QR factorization: $Q^{(k)}R^{(k)} = T^{(k)}$.

Let Q be the orthonormal matrix at convergence, then its columns are the leading eigenvectors of A, and the column space of Q gives the corresponding eigenspace.

Applying the orthogonal iterations directly to the sample covariance matrix $\mathbf{S}$ will give its sample eigenvectors which are known to be inconsistent estimators of their population counterparts when p is large. The idea of *orthogonal iterations* was used by Ma (2011) and Yang et al. (2011) to estimate the *principal subspaces* spanned by the first few, say r, leading eigen (singular) vectors for PCA and SVD, respectively. The algorithm works well when the first few vectors of the (population) PC loadings are sparse with many zeros as, for example, in the *spiked covariance model* in (1.23). In fact, for the spiked covariance model Ma (2011) has shown that the principal subspace and the leading eigenvectors can be recovered consistently and even optimally under mild conditions.

A similar approach due to Yang et al. (2011) develops an SSVD for a data matrix $\boldsymbol{X}$. Rather than estimating the individual singular vectors, their state-of-the-art approach estimates the subspaces spanned by the leading left and right singular vectors simultaneously. Consequently, these iterative methods provide sparse singular vectors that are orthogonal, a desirable property of the singular vectors lost in the sequential methods of Shen and Huang (2008), Witten et al. (2009), Lee et al. (2010), and Chen et al. (2012).

4.8 FURTHER READING

There is an interesting a growing trend in solving the sparse PCA and SVD problems without relying on optimization problems.

The adaptive sparse PCA method introduced by Johnstone and Lu (2009) does not directly involve optimization of any objective function. For large$p$, they first reduce the dimensionality from p to a smaller number k by choosing those columns of $\boldsymbol{X}$ with the highest sample variances. Once k is chosen, the standard PCA is performed on the $k \times k$ submatrix of $\boldsymbol{X}'\boldsymbol{X}$, followed by a hard-thresholding of the eigenvectors. This procedure works well when the first few vectors of the (population) PC loadings are sparse with many zeros as in the *spiked covariance model* in (1.23). The idea of thresholding coupled with orthogonal iterations are used by Ma (2011) to estimate the principal subspaces spanned by the first few leading eigenvectors. Under the spiked covariance model, the principal subspace and the leading eigenvectors can be recovered consistently, and even optimally under mild conditions. A similar approach due to Yang et al. (2011) develops an SSVD for a data matrix $\boldsymbol{X}$. Rather than estimating the individual singular vectors, the approach is designed to estimate the subspaces spanned by the leading left and right singular vectors simultaneously. Consequently, these iterative methods provide sparse singular vectors that are orthogonal, a highly desirable property lost in the sequential methods of Shen and Huang (2008), Witten et al. (2009), Lee et al. (2010) and Chen et al. (2012).

PROBLEMS

1. Prove Lemma 6.

2. Prove Lemma 7.

3. Prove the inequalities in (4.20).

4. Find the singular values and singular vectors of the matrix

$$\boldsymbol{X} = \begin{pmatrix} 2 & 1 \\ 1 & -1 \\ 1 & 1 \end{pmatrix}.$$

CHAPTER 5

SPARSE GAUSSIAN GRAPHICAL MODELS

Penalized regression methods for inducing sparsity in the *precision matrix* $\Theta = \mathbf{\Sigma}^{-1}$ have grown rapidly in the last few years. They are central to the construction of high-dimensional sparse Gaussian graphical models. In many applications, such as regression, classification, and model-based clustering, the need for computing the precision matrix $\mathbf{\Sigma}^{-1}$ is stronger than that for $\mathbf{\Sigma}$ itself. Though the former can be computed from the latter in $\mathcal{O}(p^3)$ operations, in the high-dimensional setup this could be computationally expensive and should be avoided when p is large.

The key observation is that when the random vector $\boldsymbol{Y} = (Y_1, \cdots, Y_p)'$ follows a multivariate normal distribution, the sparsity of the precision matrix relates to the notion of conditional independence of pairs of variables given the rest. More precisely, the (i, j)th entry of Θ is zero if and only if Y_i and Y_j are conditionally independent. In fact, the conditional independence of the coordinates of a Gaussian random vector $\boldsymbol{Y}$ can be displayed by a graph $G = (V, E)$, called the *Gaussian graphical model*, where V contains p vertices corresponding to the p coordinates and the edge between Y_i and Y_j is present if and only if Y_i and Y_j are not conditionally dependent. Graphical models are used extensively in genetics, social and Bayesian networks (Whittaker, 1990; Hastie et al., 2009).

Two specific examples of sparse precision matrices are presented in Section 5.1. Section 5.2, presents the connection among conditional independence, the notion of partial correlations between variables Y_i and Y_j, and the regression coefficients when regressing Y_i on all the other variables in $\boldsymbol{Y}$. A penalized likelihood method and the graphical Lasso (Glasso) algorithm for sparse estimation of the precision matrix are

High-Dimensional Covariance Estimation, First Edition. Mohsen Pourahmadi.

described in Section 5.3. The rest of the chapter describes various modifications of the Glasso estimators and their statistical properties.

5.1 COVARIANCE SELECTION MODELS: TWO EXAMPLES

The idea of estimating a covariance matrix efficiently and parsimoniously by identifying zeros in its inverse was proposed by Dempster (1972). Covariance matrices estimated in this manner are called *covariance selection models*. The motivation for this method is based on the observation that in some problems the precision matrix has a large number of zeros in its off-diagonal entries. Some earlier and specific examples are the nonstationary *ante-dependence* models of Gabriel (1962) and the stationary *autoregressive* models of Yule (1927) with banded precision matrices.

Two examples of sparse precision matrices along with their graphs are presented next.

Example 31 (Star-Shaped Models) *A sparse precision matrix given by*

$$\Theta = \begin{pmatrix} w_{11} & w_{12} & w_{13} & \cdots & w_{1p} \\ w_{12} & w_{22} & 0 & \cdots & 0 \\ w_{13} & 0 & w_{33} & \cdots & 0 \\ \vdots & \vdots & \vdots & \ddots & \vdots \\ w_{1p} & 0 & 0 & \cdots & w_{pp} \end{pmatrix}$$

is referred to as a star-shaped model due to the shape of its dependence graph (Sun and Sun, 2006), as shown in Figure 5.1.

The precision matrix of star-shaped models can be constructed by regressing any variable Y_t in the ordered vector $\boldsymbol{Y}$ on the first variable Y_1 giving the following simple model:

$$Y_1 = \varepsilon_1, Y_2 = \phi_{21}Y_1 + \varepsilon_2, \ldots, Y_p = \phi_{p1}Y_1 + \varepsilon_p.$$

Note that from (1.18) this model corresponds to setting $\phi_{tj} = 0$ for $j \geq 2$ and σ_t^2 arbitrary, and that only the first column of the T matrix in (1.20) has the nonredundant

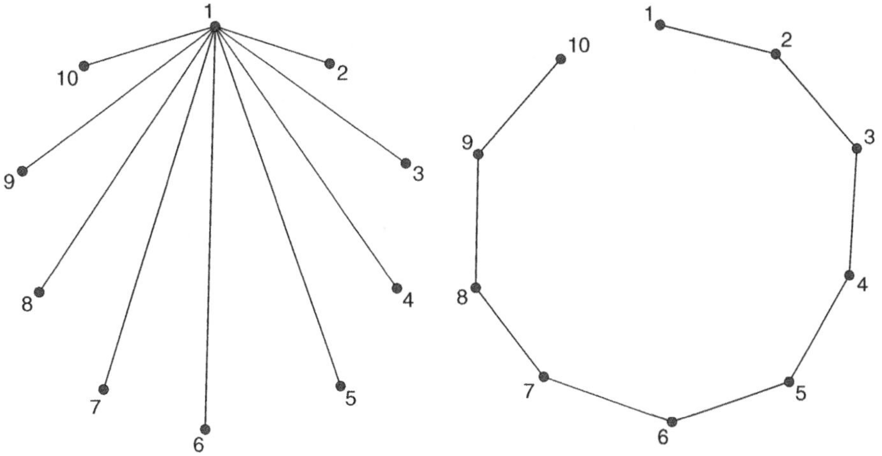

FIGURE 5.1 Graphs of star-shaped (left) and AD(1) dependence models (right).

entries $-\phi_{21}, \dots, -\phi_{p1}$. Computing $\Theta = \mathbf{\Sigma}^{-1} = T'D^{-1}T$, for this particular T and a general diagonal matrix D, it follows that

$$\Theta = \begin{pmatrix} \sigma_1^{-2} + V & -\phi_{21}\sigma_2^{-2} & -\phi_{31}\sigma_3^{-2} & \cdots & -\phi_{p1}\sigma_p^{-2} \\ -\phi_{21}\sigma_2^{-2} & \sigma_2^{-2} & 0 & \cdots & 0 \\ -\phi_{31}\sigma_3^{-2} & 0 & \sigma_3^{-2} & \cdots & 0 \\ \vdots & \vdots & \vdots & \ddots & \vdots \\ -\phi_{p1}\sigma_p^{-2} & 0 & 0 & \cdots & \sigma_p^{-2}, \end{pmatrix}$$

where $V = \sum_{j=2}^{p} \phi_{j1}^2 \sigma_j^{-2}$.

Example 32 (Ante-dependence Models) *A set of ordered random variables $Y_1, Y_2, \cdots, Y_p$ with multivariate normal distribution is said to be s-order antedependent, $AD(s)$ for short, if Y_i and Y_{i+k} are conditionally independent given the intervening variables $Y_{i+1}, Y_{i+2}, \cdots, Y_{i+k-1}$, for all i and $k \geq s$ (Gabriel, 1962). It can be shown that a p-vector $\boldsymbol{Y}$ is $AD(s)$, if and only if its precision matrix, or its modified Cholesky factor T in (1.20) is s-banded (Pourahmadi, 1999; Zimmerman, 2000). Figure 5.1 displays the graph of an ante-dependence model of order one.*

When the structure or location of zeros in the precision matrix is known, one can use the sample data $\boldsymbol{Y}_1, \cdots, \boldsymbol{Y}_n$ to estimate its nonzero entries. For example, the MLE of the nonzero entries can be found by maximizing the concentrated likelihood subject to setting a predefined subset of the entries of the precision matrix equal to zero. This equality-constrained convex optimization problem is usually solved using the iteratively proportional fitting procedure (Dempster, 1972; Whittaker, 1990). An iterative regression-based method will be presented later.

In most practical cases, it is not known which elements of the precision matrix are zero or which edges of the graph are absent, and it is plausible to try to discover these first from the data itself. The standard approach is the backward stepwise selection method. It starts by removing the least significant edges from a fully connected graph, and continues removing edges until all remaining edges are significant according to an individual partial correlation test. This procedure does not account for multiple testing; for a conservative simultaneous testing procedure, see Drton and Perlman (2004). An alternative nodewise regression method has been proposed by Meinshausen and Bühlmann (2006) for identifying the nonzero entries of Θ by fitting p *separate* regression models using each variable (node) as the response and the others as predictors. Next, we show that a regression-based interpretation of the entries of the precision matrix and the framework of penalized normal likelihood will lead to improved *joint or coupled* estimation of the p regression models.

5.2 REGRESSION INTERPRETATION OF ENTRIES OF Σ^{-1}

The regression-based interpretation of the entries of the precision matrix is the key conceptual tool behind the approach of Meinshausen and Bühlmann (2006). In this section, we give several regression interpretations of the entries of the precision matrix based on its variance–correlation decomposition:

$$\boldsymbol{\Sigma}^{-1} = (\sigma^{ij}) = DRD.$$

In particular, we show that the entries of $\boldsymbol{\Sigma}^{-1}$, R, D have direct statistical interpretations in terms of the partial correlations, regression coefficients, and variance of predicting a variable given the rest.

For this and a few other regression-based techniques, it is instructive to partition a mean zero random vector $\boldsymbol{Y}$ into two components $(\boldsymbol{Y}_1', \boldsymbol{Y}_2')'$ of dimensions p_1 and p_2, respectively, and partition its covariance and precision matrices conformally as

$$\boldsymbol{\Sigma} = \begin{pmatrix} \boldsymbol{\Sigma}_{11} & \boldsymbol{\Sigma}_{12} \\ \boldsymbol{\Sigma}_{21} & \boldsymbol{\Sigma}_{22} \end{pmatrix}, \quad \boldsymbol{\Sigma}^{-1} = \begin{pmatrix} \boldsymbol{\Sigma}^{11} & \boldsymbol{\Sigma}^{12} \\ \boldsymbol{\Sigma}^{21} & \boldsymbol{\Sigma}^{22} \end{pmatrix}.$$

Useful relationships among the blocks of $\boldsymbol{\Sigma}$ and $\boldsymbol{\Sigma}^{-1}$ are given in (3.26) and (3.27) by considering the linear least-squares regression (prediction) of $\boldsymbol{Y}_2$ based on $\boldsymbol{Y}_1$. Some special choices of $\boldsymbol{Y}_2$ corresponding to $p_2 = 1, 2$ are discussed below. They lead to powerful results connecting the regression coefficients $\Phi_{2|1}$ and the prediction error covariance $\boldsymbol{\Sigma}_{22\cdot1}$ directly to the entries of the precision matrix $\boldsymbol{\Sigma}^{-1}$.

Regression Coefficient Interpretation: Let $p_2 = 1$, $\boldsymbol{Y}_2 = Y_i$, for a fixed i, and $\boldsymbol{Y}_1 = (Y_1, \ldots, Y_{i-1}, Y_{i+1}, \ldots, Y_p)' = \boldsymbol{Y}_{-(i)}$. Then, $\boldsymbol{\Sigma}_{22\cdot1}$ is a scalar called the *partial variance* of $\boldsymbol{Y}_i$ given the rest. Define $\widehat{Y}_i$ as the linear least-squares predictor of Y_i based on the rest $\boldsymbol{Y}_{-(i)}$, and $\varepsilon_i = Y_i - \widehat{Y}_i$, $d_i^2 = \text{var}(\varepsilon_i)$ as its prediction error and prediction error variance, respectively. Then, for some β_{ij}'s

$$Y_i = \sum_{j \neq i} \beta_{ij} Y_j + \varepsilon_i, \tag{5.1}$$

and it follows immediately from (3.26) and (3.27) that the regression coefficients of Y_i on $\boldsymbol{Y}_{-(i)}$ are given by

$$\beta_{i,j} = -\frac{\sigma^{ij}}{\sigma^{ii}}, \, j \neq i, \tag{5.2}$$

and

$$d_i^2 = \text{var}(Y_i | Y_j, \, j \neq i) = \frac{1}{\sigma^{ii}}, \tag{5.3}$$

for $i = 1, \cdots, p$. This shows that σ^{ij}, the (i, j)th entry of the precision matrix is, up to a scalar, the coefficient of variable j in the multiple regression of variable i on the rest. As such, each $\beta_{i,j}$ is an unconstrained real number, however, $\beta_{j,j} = 0$ and $\beta_{i,j}$ is not symmetric in (i, j).

Matrix Inversion via Regression: It is instructive to view the formulas 3.26 and 3.27 and 5.2 and 5.3 as a way of computing the inverse of a partitioned matrix via regression. For this, fix $i = p$ in (5.2) and denote the vector of regression coefficients of Y_p on $Y_1, \cdots, Y_{p-1}$ by $\beta = (\beta_{p1}, \cdots, \beta_{p,p-1})'$. Then, the entries of the last column of the precision matrix satisfy

$$\boldsymbol{\Sigma}^{12} = -\beta\sigma^{pp}, \tag{5.4}$$

$$\sigma^{pp} = \sigma_{pp} - \beta'\boldsymbol{\Sigma}_{12}. \tag{5.5}$$

In other words, they are expressed in terms of the regression coefficients and prediction error variance. This observation is central to the development of the graphical Lasso method introduced in the next section.

An Alternative Factorization of the Precision Matrix: Rearranging and writing (5.2) in matrix form gives the following useful factorization of the precision matrix:

$$\boldsymbol{\Sigma}^{-1} = D^2(I_p - B), \tag{5.6}$$

where D is a diagonal matrix with d_j as its jth diagonal entry, and B is a $p \times p$ matrix with zeros along its diagonal and $\beta_{j,k}$ in the (j, k)th position. Now, it is evident from (5.6) that the sparsity patterns of $\boldsymbol{\Sigma}^{-1}$ and B are the same, and hence the former can be inferred from the latter using the regression setup 5.1 along with a Lasso penalty for each regression. Note that the left-hand side of (5.6) is a symmetric matrix while the right side is not necessarily so. Thus, one must impose the following *symmetry constraint*:

$$\sigma^{jj}\beta_{jk} = \sigma^{kk}\beta_{kj}, \tag{5.7}$$

for $j, k = 1, \cdots, p$.

Partial Correlation Interpretation: For $p_2 = 2$, let $\boldsymbol{Y}_2 = (Y_i, Y_j), i \neq j$ and $\boldsymbol{Y}_1 = \boldsymbol{Y}_{-(ij)}$ comprising the remaining $p - 2$ variables. Then, it follows from (3.29) that the covariance matrix between Y_i, and Y_j, after eliminating the linear effects of the other $p - 2$ components, is given by

$$\boldsymbol{\Sigma}_{22\cdot 1} = \begin{pmatrix} \sigma^{ii} & \sigma^{ij} \\ \sigma^{ij} & \sigma^{jj} \end{pmatrix}^{-1} = \Delta^{-1} \begin{pmatrix} \sigma^{jj} & -\sigma^{ij} \\ -\sigma^{ij} & \sigma^{ii} \end{pmatrix},$$

where $\Delta = \sigma^{ii}\sigma^{jj} - (\sigma^{ij})^2$. The correlation coefficient in $\boldsymbol{\Sigma}_{22\cdot 1}$ is, by definition, the *partial correlation coefficient* between Y_i and Y_j adjusted for the linear effect of others, and is given by

$$\tilde{\rho}_{ij} = -\frac{\sigma^{ij}}{\sqrt{\sigma^{ii}\sigma^{ij}}}. \tag{5.8}$$

Moreover, from (5.2) and (5.8) it follows that the regression coefficients have a *symmetric representation* in terms of the partial correlations as

$$\beta_{ij} = \tilde{\rho}_{ij}\sqrt{\frac{\sigma^{jj}}{\sigma^{ii}}}. \tag{5.9}$$

This representation, which shows that $\boldsymbol{\Sigma}^{-1}$ and $\tilde{R} = (\tilde{\rho}_{ij})$ share the same sparsity patterns, is the basis for the Peng et al., (2009) sparse partial correlation estimation algorithm. It imposes a Lasso penalty on the off-diagonal entries of the symmetric matrix of partial correlations $\tilde{R}$ and obviates the need to impose the symmetry constraint 5.7.

5.3 PENALIZED LIKELIHOOD AND GRAPHICAL LASSO

In this section, using the fact that the entries of $\boldsymbol{\Sigma}^{-1}$ have various regression-based interpretations, sparsity in the precision matrix is introduced by imposing Lasso penalties on its entries, factors, etc.

For the sample data $\boldsymbol{Y}_1, \ldots, \boldsymbol{Y}_n \sim N_p(\boldsymbol{0}, \boldsymbol{\Sigma})$, the likelihood function is

$$L(\boldsymbol{\Sigma}) = \frac{1}{(2\pi)^{np/2}|\boldsymbol{\Sigma}|^{n/2}} \exp\left\{-\frac{1}{2}\sum_{i=1}^{n} \boldsymbol{Y}_i'\boldsymbol{\Sigma}^{-1}\boldsymbol{Y}_i\right\}.$$

Let $\boldsymbol{\Theta} = \boldsymbol{\Sigma}^{-1}$ be the precision matrix and $\boldsymbol{S} = \frac{1}{n}\sum_{i=1}^{n} \boldsymbol{Y}_i \boldsymbol{Y}_i'$ the sample covariance matrix of the data. The *penalized likelihood*, up to some constants, obtained by adding to the normal log-likelihood the ℓ_1 penalty on the entries of the matrix $\boldsymbol{\Sigma}^{-1}$ is

$$\ell_P(\boldsymbol{\Theta}) = \log|\boldsymbol{\Theta}| - \mathrm{tr}(\mathbf{S}\boldsymbol{\Theta}) - \lambda||\boldsymbol{\Theta}||_1, \tag{5.10}$$

where λ is a penalty parameter. For $\lambda = 0$, the MLE of $\boldsymbol{\Sigma}$ is the sample covariance matrix $\mathbf{S}$ which is known (Ledoit and Wolf, 2004) to perform poorly when p is large and the MLE of the precision matrix may not exist when $p > n$. What is the penalized MLE of the precision matrix when $\lambda > 0$? How does one compute it?

The penalized likelihood framework for sparse graphical model estimation was proposed by Yuan and Lin (2007). They solve the problem using the interior-point method for the max log-determinant problem which guarantees the positive definiteness of the penalized MLE. This seems to have limited the applicability of the method to dimensions $p \leq 10$. A faster algorithm was proposed by Banerjee et al. (2008). To date, the fastest available algorithm is the graphical Lasso (Glasso) due to Friedman et al. (2008). It exploits effectively the connection between the Banerjee et al. (2008) blockwise interior-point procedure to recursively solve and update Lasso regression problems using the coordinate descent algorithm. A full explanation of this fruitful connection is given next following Friedman et al. (2008).

The subgradient for maximization of the penalized log-likelihood (5.10) is

$$\boldsymbol{\Theta}^{-1} - \mathbf{S} - \lambda \cdot \Gamma = 0, \tag{5.11}$$

where $\Gamma = \mathrm{Sign}(\boldsymbol{\Theta})$ and the Sign() function acts componentwise on a matrix, with $\mathrm{Sign}(\theta_{ij}) = \mathrm{sign}(\theta_{ij})$, if $\theta_{ij} \neq 0$, otherwise $\theta_{ij} \in [-1, 1]$.

From the estimating (5.11), it is not evident that the idea of regression or the Lasso algorithm could be of use in solving it. Next, it is shown that, indeed, one can recast (5.11) into the framework of linear regression so that the Lasso algorithm can be employed to compute both the covariance matrix and its inverse.

Assuming that $W = \boldsymbol{\Theta}^{-1}$ is known and viewing it as a proxy for the covariance of $\boldsymbol{Y}$, first we focus on computing the last column of $\boldsymbol{\Theta}$ as the Lasso solution of the regression coefficients β when regressing Y_p on the rest $Y_1, \cdots, Y_{p-1}$.

Let us partition W and $\mathbf{S}$ as

$$W = \begin{pmatrix} W_{11} & w_{12} \\ w_{21} & w_{22} \end{pmatrix}, \quad \mathbf{S} = \begin{pmatrix} \mathbf{S}_{11} & s_{12} \\ s_{21} & s_{22} \end{pmatrix},$$

where W_{11} and $\mathbf{S}_{11}$ are the $(p-1) \times (p-1)$ principal minors of their respective matrices and w_{22}, s_{22} are scalars. From $W\boldsymbol{\Theta} = I$, it follows that

$$\begin{aligned} W_{11}\boldsymbol{\Theta}_{12} + w_{12}\boldsymbol{\Theta}_{22} &= 0, \\ w_{12}'\boldsymbol{\Theta}_{12} + w_{22}\boldsymbol{\Theta}_{22} &= 1. \end{aligned} \tag{5.12}$$

Consider the solution of the equations arising from the upper-right block of (5.11):

$$w_{12} - s_{12} - \lambda \gamma_{12} = 0. \tag{5.13}$$

From the first identity in (5.12) it follows that

$$w_{12} = -W_{11}\boldsymbol{\Theta}_{12}/\boldsymbol{\Theta}_{22} = W_{11}\beta, \tag{5.14}$$

where $\beta = -\boldsymbol{\Theta}_{12}/\boldsymbol{\Theta}_{22}$, and

$$\text{Sign}(\beta) = -\text{Sign}(\boldsymbol{\Theta}_{12}) = -\text{Sign}(\gamma_{12}),$$

since β and $\boldsymbol{\Theta}_{12}$ have opposite signs. Substituting this and (5.14) into (5.13) leads to

$$W_{11}\beta - s_{12} + \lambda \cdot \text{Sign}(\beta) = 0, \tag{5.15}$$

which is recognized easily as the estimation equations for the Lasso regression, except that $\boldsymbol{X}'\boldsymbol{X}$ is replaced by W_{11} (see Lemma 2). Thus, in principle, β can be computed via a Lasso regression using the symbolic command

$$\text{Lasso} \quad (W_{11}, s_{12}, \lambda). \tag{5.16}$$

Now, it appears from $\boldsymbol{\Theta}_{12} = -\beta\boldsymbol{\Theta}_{22}$ that $\boldsymbol{\Theta}_{12}$ can be computed up to the scale factor $\boldsymbol{\Theta}_{22}$. However, the latter is computed from the second equation in (5.12) as

$$\boldsymbol{\Theta}_{22}^{-1} = w_{22} - w_{12}'\beta, \tag{5.17}$$

so that stacking up these two components as $(\boldsymbol{\Theta}_{12}', \boldsymbol{\Theta}_{22})'$ will give the last column of the matrix $\boldsymbol{\Theta}$. This procedure can be repeated for any other column of $\boldsymbol{\Theta}$, by first permuting its rows and columns to make the target column the last.

These developments lead to the following iterative algorithm for computing sparse precision matrices called the *graphical Lasso algorithm*. The Lasso computation in command 5.16 is done efficiently using the coordinate descent algorithm.

Graphical Lasso Algorithm:

1. Start with $W = \mathbf{S} + \lambda I$.
2. Repeat for $j = 1, 2, \cdots, p, 1, 2, \cdots, p, \cdots$ until convergence:
   (a) Partition matrix W into two parts, jth row and column, and the rest denoted symbolically by W_{11}.
   (b) Solve the estimating equations

$$W_{11}\beta - s_{12} + \lambda \cdot \text{Sign}(\beta) = 0,$$

   using the cyclical coordinate-descent algorithm to obtain $\widehat{\beta}$.
   (c) Update $w_{12} = W_{11}\widehat{\beta}$.
3. In the final cycle (for each j), solve for $\widehat{\mathbf{\Theta}}_{12} = -\widehat{\beta}\widehat{\mathbf{\Theta}}_{22}$, with $\widehat{\mathbf{\Theta}}_{22}^{-1} = w_{22} - w_{12}'\widehat{\beta}$

Some noteworthy remarks regarding the *graphical Lasso algorithm* are

1. If $W_{11} = \mathbf{S}_{11}$, then the solution $\widehat{\beta} = \text{Lasso}(\mathbf{S}_{11}, s_{12}, \lambda)$ is simply equal to the Lasso estimate of the regression coefficients of the pth variable on the remaining. It is related to the Meinshausen and Bühlmann (2006) nodewise regression approach. However, in general $W_{11} \neq \mathbf{S}_{11}$ (Banerjee et al., 2008), hence unlike the graphical Lasso algorithm, the Meinshausen and Bühlmann (2006) does not lead to the MLE of the penalized precision matrix. In fact, maximizing the penalized likelihood (5.10) amounts to more than estimating p separate Lasso regression problems, it actually solves p-coupled Lasso regressions by updating W_{11} through iterations.
2. The computational cost of the algorithm is dominated by the sparsity of $\widehat{\beta}$ and computing the update $\widehat{w}_{12} = W_{11}\widehat{\beta}$. If $\widehat{\beta}$ has r nonzero entries, the $\widehat{w}_{12}$ can be computed in rp operations, see (5.4).
3. Why is the solution positive definite?

   Note that updating the w_{12} by $\widehat{w}_{12}$ at each cycle of the Glasso algorithm, we obtain the updated

$$\tilde{W} = \begin{pmatrix} W_{11} & \widehat{w}_{12} \\ \widehat{w}_{12}' & w_{22} \end{pmatrix}.$$

   Starting with a positive-definite matrix W, we prove that the updated $\tilde{W}$ remains positive definite. From the Schur formula for determinant of partitioned matrices, we have

$$|W| = |W_{11}| \cdot (w_{22} - w_{12}' W_{11}^{-1} w_{12}).$$

Since W_{11} is already positive definite, it suffices to show that $w_{22} - w_{12}' W_{11}^{-1} w_{12} > 0$. By the updating rule 5.17, we have

$$\widehat{\Theta}_{22}^{-1} = w_{22} - \widehat{w}_{12}' W_{11}^{-1} \widehat{w}_{12} > \Theta_{22}^{-1} = w_{22} - w_{12}' W_{11}^{-1} w_{12} > 0,$$

which implies that $\tilde{W}$ is positive definite after each cycle and hence at convergence.

The Glasso and other sparse precision matrix estimation techniques exploit cross-validation to choose the tuning parameter λ. This could be expensive computationally as it requires computing the solutions over a full regularization path. A recent method proposed by Liu and Wang (2012), called tuning-insensitive graph estimation and regression (TIGER) has a *tuning-insensitive property* in the sense that it automatically adapts to the unknown sparsity pattern and is asymptotically tuning-free. In the finite sample settings, it requires minimal efforts to select the tuning or the regularization parameter. Like Glasso, it estimates the precision matrix Θ, in a column-by-column fashion where each column is computed by solving a Lasso regression problem. However, the TIGER solves this Lasso regression problem using the SQRT-Lasso (Belloni et al., 2012) where the latter also has the tuning-insensitive property. The **R** package **bigmatrix** implementing the TIGER methods is available on the Comprehensive **R** Archive Network. Under a parallel computing framework, TIGER appears to be faster and more scalable than the **Glasso** (Friedman et al., 2008) and **huge** (Zhao et al., 2012) packages in **R**.

The following toy examples illustrate the performance of the graphical Lasso algorithm in detecting a network from the sample data. The computation is done using the **Glasso** package in **R**.

Example 33 (Star-Shaped Models) *We simulated $n = 1000$ observations from a multivariate normal distribution with $p = 10$, zero mean vector, and a star-shaped covariance matrix introduced in Example 31. The estimated graphs corresponding to the sparse precision matrices for different values of the penalty $\lambda = 0.5, 0.1$ are displayed in Figure 5.2. Note that as the penalty parameter increases, the graph becomes more sparse with fewer edges. The plots in the left panels correspond to the graphical Lasso solutions, and those in right correspond to the Meinshausen and Bühlmann (2006) method (MB) based on p separate nodewise Lasso regressions.*

Example 34 (Ante-Dependence Models) *We simulated $n = 1000$ samples from a multivariate normal distribution with $p = 10$, zero mean vector, and an AD(1) covariance matrix. Figure 5.1 shows the network plot for the true precision matrix of an AD(1) model. The four graphs for the precision matrix corresponding to different values of the penalty λ are displayed in Figure 5.3. Note that as the penalty parameter increases, the graph becomes more sparse with fewer edges. The Meinshausen and Bühlmann (2006) method (MB) is also used and the output is displayed in Figure 5.3.*

Example 35 (Pitprops Data) *Recall the Pitprops data from Chapter 4 with $p = 13$ variables. The ℓ_1 penalty in (5.10) would introduce sparsity into Θ; the larger the λ, the more sparse the estimate of Θ. We solve the optimization problem using the*

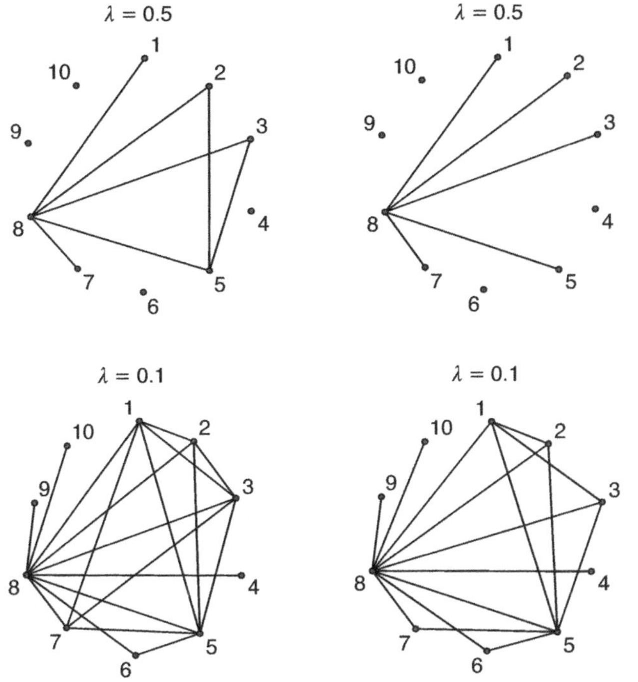

FIGURE 5.2 Graphical Lasso (left) and MB (right) solutions for star-shaped model.

graphical Lasso algorithm described previouly and implemented in the R package. The four graphs corresponding to four different values of the penalty λ are displayed in Figure 5.4. Note that as the penalty parameter increases, the graph becomes more sparse with fewer edges.

5.4 PENALIZED QUASI-LIKELIHOOD FORMULATION

There are several approximate likelihood methods in the literature designed to overcome some of the shortcomings of the approach of Meinshausen and Bühlmann (2006) like its lack of symmetry.

The sparse pseudo-likelihood inverse covariance estimation (SPLICE) algorithm of Rocha et al. (2008) and the SPACE algorithm of Peng et al. (2009) also impose sparsity constraints directly on the precision matrix, but with slightly different regression-based reparameterizations of $\boldsymbol{\Sigma}^{-1}$ (see (5.6) and (5.8)). They are designed to improve the approach of Meinshausen and Bühlmann (2006) including its lack of symmetry for neighborhood selection in Gaussian graphical models. While Meinshausen and Bühlmann (2006) use p separate linear regressions to estimate the neighborhood of one node at a time, Rocha et al. (2008) and Peng et al. (2009) propose merging all p linear regressions into a single least-squares problem where the observations associated to each regression are weighted differently according to their conditional variances.

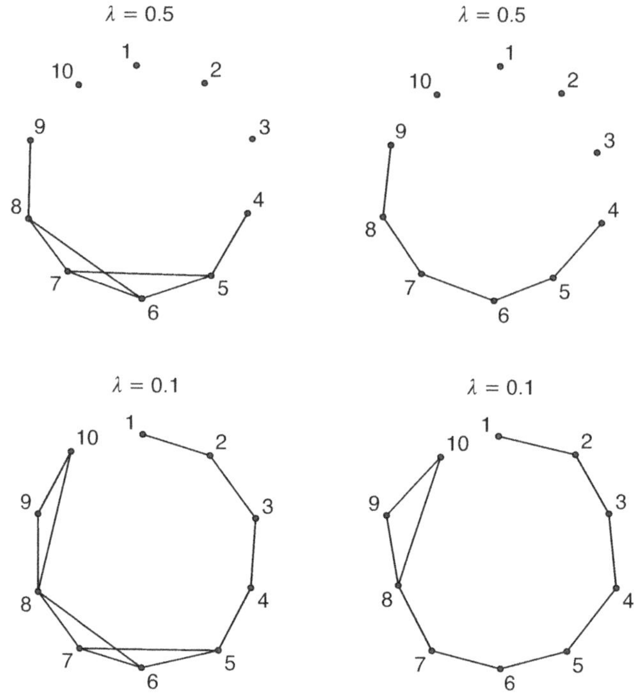

FIGURE 5.3 Graphical Lasso (left) and MB (right) solutions for the AD(1).

To appreciate the need for approximate or pseudo-likelihood, it is instructive to note that unlike the sequence of prediction errors in (1.18), the ε_j's from Section 5.2 are correlated so that D^2 is not really the covariance matrix of the vector of regression errors $\varepsilon = (\varepsilon_1, \cdots, \varepsilon_p)'$. The use of its true covariance matrix in the normal log-likelihood would increase the computational cost at the estimation stage. This problem is circumvented in Rocha et al. (2008) and Friedman et al. (2010) by using a pseudo-likelihood function which in the normal case amounts to using D^2 instead of $Cov(\varepsilon)$. To this pseudo-log-likelihood function, they add the symmetry constraints (5.7) and a weighted Lasso penalty on the off-diagonal entries to promote sparsity.

A drawback of both SPLICE and SPACE algorithms is that they do not enforce the positive-definiteness constraint, hence the resulting covariance estimator is not guaranteed to be positive definite.

5.5 PENALIZING THE CHOLESKY FACTOR

In this section, for *ordered* variables in a random vector we take advantage of the regression interpretation of the entries of the Cholesky factor of the precision matrix (1.20) and induce sparsity in T through the penalized likelihood.

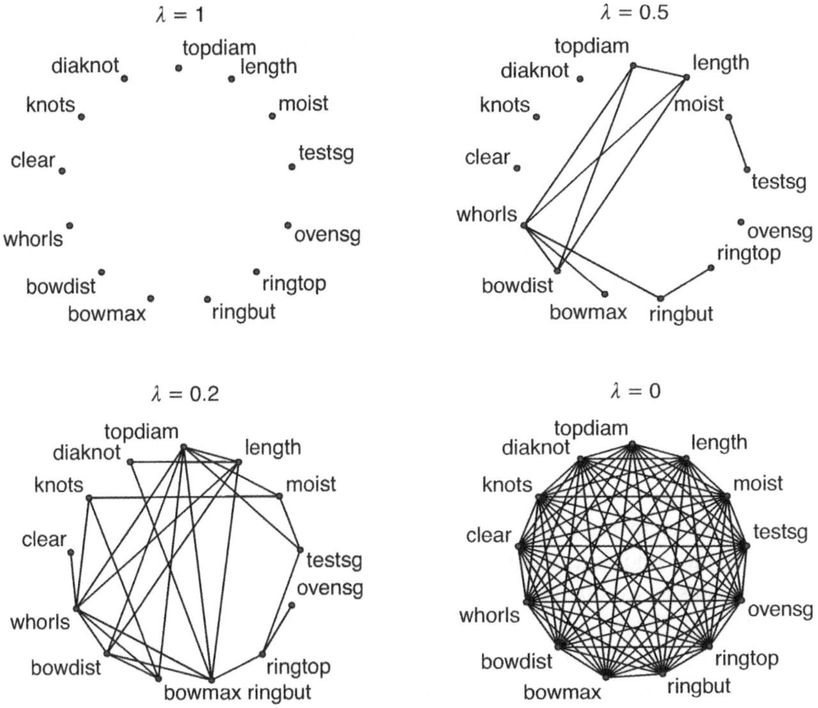

FIGURE 5.4 Four different graphical Lasso solutions for the Pitprops data.

In one of the first attempts at inducing sparsity in the precision matrix, Huang et al. (2006) proposed adding to the negative log-likelihood a Lasso penalty on the off-diagonal entries of the Cholesky factor. An sparse T is then found by minimizing

$$n \log |\widehat{\boldsymbol{\Sigma}}| + \sum_{i=1}^{n} \boldsymbol{Y}_i' \widehat{T} \widehat{D}^{-1} \widehat{T} \boldsymbol{Y}_i + \lambda \sum_{t=2}^{p} P(\phi_t), \tag{5.18}$$

$$P(\phi_t) = \sum_{j=1}^{t-1} |\phi_{t,j}|, \tag{5.19}$$

where $\phi_t = (\phi_{t1}, \cdots, \phi_{t,t-1})'$ is the tth row of the Cholesky factor T or the vector of regression coefficients when Y_t is regressed on its predecessors.

Using a local quadratic approximation to the penalty function, it was shown that the method is computationally tractable (see also Huang et al., 2007). The ℓ_1 penalty usually results in zeros placed in T with no regular patterns so that the sparsity of the precision matrix is not guaranteed.

Fortunately, the patterns of zeros in T can be controlled much better and pushed farther away from the main diagonal using a *nested* Lasso penalty proposed by Levina et al. (2008). They replaced the sum $\sum_{j=1}^{t-1} |\phi_{t,j}|$ in (5.19) by

$$P(\phi_t) = |\phi_{t,t-1}| + \frac{|\phi_{t,t-2}|}{|\phi_{t,t-1}|} + \cdots + \frac{|\phi_{t,1}|}{|\phi_{t,2}|}, \tag{5.20}$$

where $0/0$ is defined to be zero. The effect of such penalty is to have a contiguous pattern of zeros in T right below the main diagonal, in the sense that if the jth variable is not included in the tth regression ($\phi_{tj} = 0$), then all the preceding variables are also excluded, since giving them nonzero coefficients would result in an infinite penalty. Hence, the tth regression only uses the k_t immediate predecessors of the tth variable, and each regression could have a different order k_t . The scaling of coefficients could be an issue; see Levina et al. (2008) for various alternatives. Since this formulation is highly nonconvex and nonlinear, an iterative local quadratic approximation algorithm was used to fit the model. When all the regressions have the same order ($k_t = k$), then the Cholesky factor T is said to be k-banded. In this case, Bickel and Levina (2008a) provide conditions ensuring consistency in the operator norm for precision matrix estimates based on banded Cholesky factors.

Motivated by the nested banding structure, Leng and Li (2011) have proposed a novel *forward adaptive banding* approach for estimating T using variable band lengths $\{k_j; j = 1, \cdots, p\}$. These are estimated using the modified Bayes information criterion (BIC):

$$\text{BIC} = n \log |\widehat{\Sigma}| + \sum_{i=1}^{n} Y_i' \widehat{T} \widehat{D}^{-1} \widehat{T} Y_i + C_n \log(n) \sum_{j=1}^{p} k_j, \tag{5.21}$$

where $\widehat{T}$ and $\widehat{D}$ are based on estimated autoregressive coefficients and prediction variances obtained by fitting AR(k_j) models to the successive measurements. The k_j's are chosen by minimizing (5.21) over

$$k_j \leq \min\{n/(\log n)^2, j-1\}, \quad j = 1, \cdots, p,$$

with a sequence C_n diverging to infinity. Note that (5.21) can be decoupled as

$$\text{BIC} = \sum_{j=1}^{p} \left\{ n \log \widehat{\sigma}_j^2 + \sum_{i=1}^{n} \widehat{\varepsilon}_{ij}^2 / \widehat{\sigma}_j^2 + C_n \log(n) k_j \right\},$$

so that each k_j can be chosen separately.

Under the regularity conditions stated below, consistency of the order selection for the autoregressive models and rates of convergence for covariance and precision matrix estimators were established by Leng and Li (2011).

Assumption 0 Sparsity: There exists a sequence of integers $k_{0,j} < j, j = 2, \cdots, p$, such that the autoregressive coefficients satisfy

$$\phi_{j,j-\ell} = 0, \quad \text{for all} \quad k_{0,j} < \ell < j \quad \text{and} \quad \phi_{j,j-k_{0,j}} \neq 0.$$

This assumption implies that variables far apart are conditionally independent when $\boldsymbol{Y}$ is normally distributed and hence implies sparsity of the precision matrix. Note that the choice of $k_{0,j} = j - 1$ imposes no constraints on the matrix T.

Assumption 1 Bounded Eigenvalues: There exists a $\kappa > 0$ such that for all $j = 2, \cdots, p$,

$$\lambda_{\min}\{\text{cov}(Y_j, \cdots, Y_{j-k_{0,j}})\} > \kappa,$$

where $\lambda_{\min}(\cdot)$ stands for the smallest eigenvalue of the relevant matrix.

Assumption 2 $C_n \to \infty$, and for all $j = 2, \cdots, p$,

$$|\phi_{j,j-k_{0,j}}|(n/C_n k_{0,j} \log n)^{1/2} \to \infty \quad \text{and} \quad C_n k_{0,j} \log n/n \to 0.$$

Assumption 3 $p = O(n^\gamma)$ for some $\gamma > 0$.

Theorem 11 *Let $k_{max} = \max_{1 \leq j \leq p} k_{0,j}$. Under Assumptions 0–3,*

(a) the modified BIC is model selection consistent, that is, as $n \to \infty$

$$P(\widehat{k}_j = k_{0,j}, j = 2, \cdots, p) \to 1.$$

(b) if $k_{max}(\log p/n)^{1/2} \to 0$, the estimated covariance and precision matrices satisfy the following rate of convergence in the operator norm with probability tending to 1,

$$||\widehat{\boldsymbol{\Sigma}}^{-1} - \boldsymbol{\Sigma}^{-1}|| = O_p(k_{max}(\log p/n)^{1/2}) = ||\widehat{\boldsymbol{\Sigma}} - \boldsymbol{\Sigma}||.$$

The rate of convergence in part (b) of the theorem parallels Theorem 3 in Bickel and Levina (2008a).

Chang and Tsay (2010) have extended the Huang et al., (2006) setup using an equi-angular penalty which imposes different penalty on each regression or row of T. The penalties are inversely proportional to the prediction error variance σ_t^2

of the tth regression. They use extensive simulations to compare the performance of their method with others including the sample covariance matrix, the banding (Bickel and Levina, 2008a)), and the ℓ_1-penalized normal log-likelihood (Huang et al., 2006). Contrary to the banding method, the method of Huang et al. (2006) and the equi-angular method work reasonably well for six covariance matrices with the equi-angular method outperforming the others. Since the modified Cholesky decomposition is not permutation-invariant, they also use a random permutation of the variables before estimation to study sensitivity to permutation of each method. They conclude that permuting the variables introduces some difficulties for each estimation method, except the sample covariance matrix, but the equi-angular method remains the best with the banding method having the worst sensitivity to permutation. They also compare these methods by applying them to a portfolio selection problem with $p = 80$ series of actual daily stock returns.

A regression interpretation of the Cholesky factor of the covariance matrix, as opposed to the more standard regression interpretation of the Cholesky factor of the inverse covariance is given by Rothman et al. (2010a) and Pourahmadi (2007a). Regularizing this Cholesky factor in view of its regression interpretation always results in a positive-definite estimator. In particular, one obtains a positive-definite banded estimator of the covariance matrix at the same computational cost as the banded estimator of Bickel and Levina (2008a), where the latter is not guaranteed to be positive definite. Rothman et al. (2010a) establish theoretical connections between banding Cholesky factors of the covariance matrix and its inverse and maximum likelihood estimation under the banding constraint. A more direct method of penalized covariance estimation is given by Lam and Fan (2009) and Bien and Tibshirani (2011).

In the context of space–time data, Zhu and Liu (2009) rely on the Cholesky decomposition of the precision matrix, based on several ordering schemes using the spatial locations of the observations. Their theoretical and simulation studies show that the regression-based penalized normal likelihood method performs competitively.

5.6 CONSISTENCY AND SPARSISTENCY

Some theoretical properties of the ℓ_1-penalized normal likelihood estimator of the precision matrix in the large p scenario are discussed in this section. *Sparsistency* refers to the property that all zero entries are actually estimated as zero with probability tending to 1.

Rothman et al. (2008) showed that the rate of convergence of the ℓ_1-penalized normal likelihood estimator of the precision matrix in the Frobenius norm is of the order $(s \log p/n)^{1/2}$, where $s = s_n$ is the total number of nonzero elements in the precision matrix. This reveals that the contribution of high-dimensionality is merely of a logarithmic factor. Fan et al. (2009) and Lam and Fan (2009) extended this penalized likelihood approach to general nonconvex penalties, such as the SCAD (Fan and Li, 2001). The pseudo-likelihood method proposed by Peng et al. (2009) enjoys consistency in terms of both estimation and model selection.

The *sparsistency* and rates of convergence for sparse covariance and precision matrix estimation using the penalized likelihood with nonconvex penalty functions have been studied in Lam and Fan (2009). In a given situation, sparsity might be present in the covariance matrix, its inverse, or Cholesky factor. They develop a unified framework to study these three sparsity problems with a general penalty function and show that the rates of convergence for these problems under the Frobenius norm are of order $(s \log p/n)^{1/2}$, where $s = s_n$ is the number of nonzero elements, $p = p_n$ is the size of the covariance matrix, and n is the sample size.

In addition to sparsity of the precision matrix discussed so far, it is possible to study sparse estimation of the covariance matrix itself where one is interested in finding a covariance matrix $\boldsymbol{\Sigma}$ under which the variables are more likely to be marginally independent. Lam and Fan (2009) proposed using a local minimizer of the objective function

$$\ell_P(\boldsymbol{\Sigma}) = -\log|\boldsymbol{\Sigma}| + \mathrm{tr}(\mathbf{S}\boldsymbol{\Sigma}^{-1}) + \lambda \sum_{i \neq j} |\sigma_{ij}|, \tag{5.22}$$

where λ is a penalty parameter. They established a convergence rate bound for such estimator similar to that for the precision matrix discussed before, but no algorithm for its computation was developed. Since the above penalized likelihood function is not convex, recently Bien and Tibshirani (2011) have proposed a majorize–minimize approach in which they iteratively solve convex approximations to the original non-convex problem.

5.7 JOINT GRAPHICAL MODELS

In this section, we present a joint graphical Lasso technique for jointly estimating multiple graphical models corresponding to distinct but related populations such as cancerous and normal tissues. It is a natural extension of the graphical Lasso algorithm to the case of multiple data sets using penalized log-likelihood approach, where the choice of the penalty function depends on the characteristics of the graphs that we expect to be shared across various conditions.

The Gaussian graphical model discussed so far is based on data from a single population where it is assumed that each observation is drawn from the same Gaussian distribution. However, observations could come from several distinct populations. For instance, consider collecting gene expression measurements from samples of cancerous tissue and a set of healthy tissue samples. In this case, one might want to estimate graphical models for the cancer and the healthy samples. It is expected that the two graphical models to be similar to each other, since both are based upon the same type of tissue, but also to have important differences. Estimating separate graphical models for the cancer and normal samples does not exploit the potential similarity between the true graphical models, and estimating a single graphical model for the cancer and normal samples ignores the fact that we do not expect the true graphical models to be identical, and that the differences between the graphical models may be of interest.

Suppose that there are independent sample data from K normal populations with mean zero and $p \times p$ covariance and precision matrices $\mathbf{\Sigma}^{(k)}, \mathbf{\Theta}^{(k)}, k = 1 \cdots, K$. It is assumed that p common variables are measured across the populations, the sample sizes are $n_k, k = 1, \cdots, K$ and the sample covariance matrices are denoted by $\mathbf{S}_k$. Let $\mathbf{\Theta}$ stand for the collection of all K precision matrices, then the penalized log-likelihood for the data is of the form

$$\ell_P(\mathbf{\Theta}) = \sum_{k=1}^{K} n_k \left(\log|\mathbf{\Theta}^{(k)}| - \mathrm{tr}(\mathbf{S}^{(k)}\mathbf{\Theta}^{(k)})\right) - P(\mathbf{\Theta}), \tag{5.23}$$

where $P(\cdot)$ is a convex penalty function to be discussed later. Note that when $P(\mathbf{\Theta}) = 0$, then maximizing (5.23) amounts to estimating K graphical models or covariances separately and the MLE of $\mathbf{\Sigma}^{(k)}$ is the sample covariance matrix $\mathbf{S}^{(k)}$ provided that p is small. What are the MLE of the precision matrices when they have certain common characteristics or when p is bigger than at least one of the n_k's?

A group Lasso penalty of the form

$$P(\mathbf{\Theta}) = \lambda \sum_{i \neq j} \left(\sum_k |\theta_{ij}^{(k)}|\right)^{1/2} \tag{5.24}$$

was proposed by Guo et al. (2011) which forces similar patterns of sparsity across the K populations. This function is not convex, nevertheless they compute the solution of (5.23) using an iterative approach based on local linear approximation of the nonconvex objective function. In particular, let $(\theta_{ij}^{(k)})_{(m)}$ be the estimate of $\theta_{ij}^{(k)}$ at the mth iteration, then using a local linear approximation one may write

$$\sqrt{\sum_{k=1}^{K} |\theta_{ij}^{(k)}|} = \frac{\sum_{k=1}^{K} |\theta_{ij}^{(k)}|}{\sqrt{\sum_{k=1}^{K} |(\theta_{ij}^{(k)})_{(m)}|}} \equiv \alpha_{ij}^{(m)} \sum_{k=1}^{K} |\theta_{ij}^{(k)}|, \tag{5.25}$$

where $\alpha_{ij}^{(m)} = 1/\sqrt{\sum_{k=1}^{K} |(\theta_{ij}^{(k)})_{(m)}|}$. Interestingly, this approximation decouples (5.23) into K separate optimization problems where the graphical Lasso algorithm can be used to find their solutions iteratively. The asymptotic properties of the estimators from this joint estimation method, including consistency, as well as sparsistency are established, when both p and n go to infinity and the tuning parameter goes to zero at a certain rate, under certain regularity conditions on the true precision matrices (for more details see Guo et al., 2011).

Ideally, the nonzero penalty function $P(\cdot)$ should be chosen so that the $\mathbf{\Theta}^{(k)}$'s are encouraged to be sparse and share some common characteristics like the location of nonzero entries, etc. We consider the penalty function

$$P(\mathbf{\Theta}) = Q(\mathbf{\Theta}) + \lambda_1 \sum_k \sum_{i \neq j} |\theta_{ij}^{(k)}|, \tag{5.26}$$

where $Q(\cdot)$ a convex penalty function will be chosen to encourage shared properties across the K precision matrices, and λ_1 is a tuning parameter. Note that when $Q = 0$, then (5.26) amounts to performing K decoupled graphical Lasso algorithms. Hence in general, the solution of the optimization problem (5.26) is referred to as the *joint graphical Lasso* (Danaher et al., 2011) solution.

The following two choices of Q are of particular interest:

$$Q_{\text{FGL}}(\mathbf{\Theta}) = \lambda_2 \sum_{k<k'} \sum_{i,j} |\theta_{ij}^{(k)} - \theta_{ij}^{(k')}|, \; Q_{\text{GGL}}(\mathbf{\Theta}) = \lambda_2 \sum_{i \neq j} \left(\sum_{k=1}^{K} \theta_{ij}^{(k)^2} \right)^{1/2}, \tag{5.27}$$

where the subscripts FGL and GGL stand for *fused graphical Lasso* and *group graphical Lasso*, respectively.

For large λ_2, it is evident that the FGL penalty encourages many elements of the $\mathbf{\Theta}^{(k)}$'s to be *identical* across the K populations. It borrows information across the populations leading to similar network structure with similar edge values. In contrast, the group Lasso penalty encourages a similar pattern of sparsity across all of the precision matrices. It has the tendency to create *zeros* in the same places in the K estimated precision matrices. A generalized gradient decent algorithm for maximizing the objective functions 5.26 with Q's as in (5.27) is given by Danaher et al. (2011). Their **R** package **JGL** performs the joint graphical Lasso for sparse estimation of several precision matrices corresponding to the two penalty functions.

5.8 FURTHER READING

The Glasso or estimating a precision matrix by performing penalized maximum likelihood estimation has been the subject of intense study in the last few years. There is a vast literature on computational and statistical properties of the estimators. Theoretical properties of the penalized likelihood methods are studied by Rothman et al. (2008), Lam and Fan (2009), Ravikumar et al. (2011). Yuan (2010) and Cai et al. (2011) have proposed the graphical Dantzig selector and constrained L_1-minimization for inverse matrix estimation (CLIME), respectively, which can be solved by linear programming and appear to be more amenable to theoretical analysis than the penalized likelihood method. We note that CLIME is a non-Lasso type estimator solving the optimization problem

$$\text{minimize } ||\mathbf{\Theta}||_1 \quad \text{subject to} \quad ||\boldsymbol{S}\mathbf{\Theta} - \boldsymbol{I}||_\infty \leq \lambda,$$

where λ is the tuning parameter.

Sparse estimation of a precision matrix was originally formulated as log-determinant (log-det) semidefinite programming (SDP) problems with a large number of linear constraints (Yuan and Lin, 2007; Ravikumar et al., 2011). Standard primal-dual interior-point methods based on solving the Schur complement equation would encounter severe computational bottlenecks if they are applied to solve these SDPs. For a customized inexact primal-dual path-following interior-point algorithm for solving large-scale log-det SDP problems arising from sparse precision matrix estimation, see Li and Toh (2010) and references therein.

PROBLEMS

1. **(a)** Let $\boldsymbol{\Sigma}$ be the $p \times p$ covariance matrix of a star-shaped model in Example 31 and $\boldsymbol{u}$ be a fixed p-vector. Compute the precision matrix of $\boldsymbol{\Sigma} + \boldsymbol{u}\boldsymbol{u}'$ when it exists and discuss the nature of its sparsity as it relates to the sparsity of the vector $\boldsymbol{u}$.

   (b) Repeat (a) for the AD(s) covariance model in Example 32.

2. For the 3×3 covariance matrix

$$\boldsymbol{\Sigma} = \begin{pmatrix} 1 & 0.9 & 0.7 \\ 0.9 & 1 & 0.4 \\ 0.7 & 0.4 & 1 \end{pmatrix},$$

   compute the matrices B, D and $\tilde{R}$, introduced in Section 5.2.

3. For the 5×5 covariance matrix in Example 17(b) compute the matrices B, D, and $\tilde{R}$ introduced in Section 5.2.

4. Compute the matrices B, D, and $\tilde{R}$ introduced in Section 5.2 for the following matrices:

   (a) the star-shaped covariance model,

   (b) the compound symmetry covariance matrix,

   (c) the AR(1) covariance model,

   (d) the AD(1) covariance model,

   (e) the MA(1) covariance model.

5. Let $\boldsymbol{\Theta}$ be the $p \times p$ precision matrix of a star-shaped model in Example 31 with $w_{ii} = 1, i = 1, \ldots, p$. Find the range of values for $w_{1j}, j = 2, \ldots, p$ so that $\boldsymbol{\Theta}$ is positive definite.

CHAPTER 6

BANDING, TAPERING, AND THRESHOLDING

Heuristically, very large covariance matrices are bound to have very many zeros or small entries. In other words, they are *sparse*, and componentwise regularization or estimating small entries by zeros seems like a natural thing to do. Interestingly, experience in the context of classification in machine learning (Domingos and Pazzani, 1997) and microarray data (Dudoit et al., 2002) suggests that using a diagonal covariance matrix or ignoring the off-diagonal entries can lead to better classification rules than those based on estimating the whole covariance matrix. To study and explain this empirical phenomenon, it is helpful to work with a class of covariance structures spanning the whole range between independence and arbitrary dependence structures. Then, by relying on the data one can decide whether to estimate or ignore certain subdiagonals.

Banding is a simple and systematic way of estimating large covariance matrices, it starts estimating the covariance matrix by a diagonal matrix and then successively adds the other subdiagonals by estimating the first, second, $\cdots$, ℓth subdiagonals, if warranted by the data or the application area. The key question in implementing banding then is: how to choose the tuning or the *banding* parameter ℓ? Tapering is a smooth version of banding where it gradually shrinks the off-diagonal entries toward zero. Banding and tapering require a natural ordering among the variables and assume that variables farther apart in the ordering are less correlated. These approaches are appropriate for time series, longitudinal and spectroscopic data, or in situations where there is a metric on the variable indexes. In applications where the variables are not ordered, like gene expression arrays and stocks in a portfolio, covariance estimators are required to be invariant under variable permutations. Thresholding

High-Dimensional Covariance Estimation, First Edition. Mohsen Pourahmadi.

the sample covariance matrix does have this property. Compared to banding, the pattern of zeros in a thresholded covariance matrix is fairly irregular. Choosing the threshold parameter λ is the key in implementing it. Since the entries of the sample covariance matrix have different variability, adaptive thresholding where each entry has a different threshold parameter seems more appropriate in practice.

The topic of componentwise regularization of the sample covariance matrix studied in this chapter has the goal of replacing the smaller entries of $\mathbf{S}$ by zero using either banding or thresholding. Unlike the Ledoit–Wolf estimator which shrinks only the eigenvalues, these methods affect both the eigenvalues and eigenvectors of the covariance estimator. Conditions for high-dimensional consistency of such estimators will be given and convergence of $\log p/n$ to zero plays a central role. These methods are computationally simple, indeed, the only computational demand is the selection of a tuning parameter.

6.1 BANDING THE SAMPLE COVARIANCE MATRIX

In this section, estimation of a covariance matrix via banding is presented. Since for high-dimensional data the sample covariance matrix $\mathbf{S}$ is singular, banding is a way to obtain a well-conditioned and nonsingular estimator.

Given a $p \times p$ sample covariance matrix $\mathbf{S} = (s_{ij})$ and any integer ℓ, $0 \leq \ell \leq p$, its ℓ-banded (Bickel and Levina, 2008a) version defined by

$$B_\ell(\mathbf{S}) = [s_{ij}\mathbf{1}(|i-j| \leq \ell)],$$

can serve as an estimator for Σ. This regularization is ideal when the variables or their indices can be arranged so that entries of the covariance matrix farther away from the main diagonal are negligible:

$$|i-j| > k \Longrightarrow \sigma_{ij} = 0.$$

This occurs, for example, if $\mathbf{\Sigma}$ is the covariance matrix of $\boldsymbol{Y} = (Y_1, \cdots, Y_p)'$ where $Y_1, Y_2, \cdots$, defined by the moving average process

$$Y_t = \sum_{j=1}^{k} \theta_{t,t-j}\varepsilon_j,$$

where ε_j's are i.i.d. with mean 0 and finite variances.

The performance of the estimator $\mathbf{S}_{\ell,p} = B_\ell(\mathbf{S})$ depends critically on the optimal choice of the banding parameter ℓ which is usually done using a cross-validation method. The consistency in operator norm is important in applications as it implies the consistency of the eigenvalues and eigenvectors, and hence the PCA (Johnstone and Lu, 2009), and related eigenstructure-based methods in multivariate statistics when n is small and p is large, see Theorem 5.

Asymptotic analysis of banded estimators when n, p, and ℓ are large can be studied for a class of *bandable* covariance matrices defined next. Consider the set of well-conditioned covariance matrices $\mathbf{\Sigma}_p$ such that for all p,

$$\mathcal{C}(\varepsilon) = \left\{\mathbf{\Sigma}_p : 0 < \varepsilon \leq \lambda_{min}(\mathbf{\Sigma}_p) \leq \lambda_{max}(\mathbf{\Sigma}_p) \leq \varepsilon^{-1}\right\},$$

where $\varepsilon > 0$ is fixed and independent of p. Now, for a given $\alpha > 0$, consider the following class of bandable (sparse) positive-definite matrices:

$$\mathcal{U}(\alpha, \varepsilon) \quad = \left\{\mathbf{\Sigma} \in \mathcal{C}(\varepsilon) \; : \; \max_j \textstyle\sum_i\{|\sigma_{ij}|; \; |i-j| > k\} \leq Ck^{-\alpha}\right\}, \qquad (6.1)$$

where $C > 0$. Note that the parameter α controls the rate of decay of the covariance entries σ_{ij} as one moves away from the main diagonal. The optimal rate of convergence of estimating a covariance matrix from this bandable class depends critically on α. The following theorem due to Bickel and Levina (2008a) shows that the banded estimator is consistent in the operator (spectral) norm, uniformly over the class of approximately "bandable" matrices in (6.1), provided that $\log p/n \to 0$. Moreover, they obtain explicit rate of convergence which depends on how fast the banding parameter $\ell \to \infty$.

Theorem 12 *Suppose* $\mathbf{Y}$ *is Gaussian with* $\mathbf{\Sigma} \in \mathcal{U}(\alpha, \varepsilon_0)$ *for an* $\alpha > 0$. *If* $\ell = \ell_n \sim \left(\frac{n}{\log p}\right)^{\frac{1}{2(\alpha+1)}}$, *then*

(a) $||\mathbf{S}_{\ell,p} - \mathbf{\Sigma}_p|| = O_P\left(\left(\frac{\log p}{n}\right)^{\frac{\alpha}{2(\alpha+1)}}\right)$, *uniformly in* $\mathbf{\Sigma} \in \mathcal{U}$,

(b) the same holds for $||\mathbf{S}_{\ell,p}^{-1} - \mathbf{\Sigma}_p^{-1}||$.

Note that the optimal band ℓ_n depends on n and p, as well as the α which controls the dependence and its rate of decay. Theorem 1(a) implies that $||\mathbf{S}_{\ell,p} - \mathbf{\Sigma}_p|| \to 0$ uniformly if $\log p/n \to 0$.

An ℓ-banded matrix $B_\ell(\mathbf{S})$ is not necessarily positive definite. The tapering operation introduced in the next section is useful in guaranteeing the positive-definiteness in some cases. The idea of banding and regularizing the lower triangular matrix of the Cholesky decomposition of $\mathbf{\Sigma}^{-1}$ has been studied in Wu and Pourahmadi (2003), Huang et al. (2006), and Bickel and Levina (2008a) which is effective in guaranteeing the positive-definiteness of the estimated covariance matrix. A consistency result and rate of convergence similar to that in Theorem 12 are given in Bickel and Levina (2008a) for the banded Cholesky factor.

6.2 TAPERING THE SAMPLE COVARIANCE MATRIX

Tapering the covariance matrix has a long history in time series analysis, it has been used recently to improve the performance of linear discriminant analysis (Bickel and Levina, 2004) and Kalman filter ensembles (Furrer and Bengtsson, 2007).

A tapered estimator of the sample covariance matrix for a tapering matrix $W = (w_{ij})$ replaces $\mathbf{S}$ by

$$\mathbf{S}_W = \mathbf{S} * W = (s_{ij} w_{ij}),$$

where $(*)$ denotes the Schur (coordinate-wise) matrix multiplication. When $W = (w_{ij})$ is a positive-definite symmetric matrix, then $\mathbf{S}_W$ as the Schur product of two positive-definite matrices is guaranteed to be positive definite (see Theorem 2). We note that banding corresponds to the *rectangular* weight $W = (w_{ij}) = (\mathbf{1}(|i - j| \leq \ell)$ which is not a positive-definite matrix. The choice of a smoother positive-definite tapering matrix W with off-diagonal entries gradually decaying to zero will ensure the positive-definiteness as well as optimal rate of convergence of the tapered estimator. For example, the *trapezoidal* weight matrix given by

$$w_{ij} = \begin{cases} 1, & \text{if } |i - j| \leq \ell_h, \\ 2 - \frac{|i-j|}{\ell_h}, & \text{if } \ell_h < |i - j| < \ell, \\ 0, & \text{otherwise,} \end{cases}$$

for a given tapering parameter ℓ with $\ell_h = \ell/2$ is a popular taper in the literature of time series analysis.

In the tapering context, it is possible to consider a larger class of covariance matrices than $\mathcal{U}(\alpha, \varepsilon)$ where their smallest eigenvalue is allowed to be zero:

$$\mathcal{T}(\alpha, \varepsilon) = \left\{ \mathbf{\Sigma} : \max_j \textstyle\sum_i \{|\sigma_{ij}|;\ |i - j| > k\} \leq Ck^{-\alpha}, \lambda_{max}(\mathbf{\Sigma}) \leq \varepsilon^{-1} \right\}. \tag{6.2}$$

Note that as in (6.1) the parameter α controls the rate of decay of the covariance entries σ_{ij} as one moves away from the main diagonal. The optimal rate of convergence for tapered estimator under both the operator and Frobenius norms are derived by (Cai et al., 2010). They also carry out a simulation study to compare the finite sample performance of their proposed estimator with that of the banding estimator introduced in Bickel and Levina (2008a). The simulation shows that the tapered estimator has good numerical performance, and it nearly uniformly outperforms the banding estimator. However, the proposed tapering does not improve the bound under the Frobenius norm, so that under this norm banding performs as good as tapering.

The techniques of banding and tapering have been used in the context of time series analysis where one usually has only one ($n = 1$) realization (Wu and Pourahmadi, 2009; McMurray and Politis, 2010; Bickel and Gel, 2011).

6.3 THRESHOLDING THE SAMPLE COVARIANCE MATRIX

In high dimensions, it is plausible that many elements of the population covariance matrix could be small and hence $\mathbf{\Sigma}$ could be sparse. How does one develop an estimator other than $\mathbf{S}$ that can cope with this additional information? The technique of thresholding, originally developed in nonparametric function estimation, has been used in the estimation of large covariance matrices by Bickel and Levina (2008b), El Karoui (2008), and Rothman et al. (2009). Unlike banding and tapering, thresholding does not require the variables to be ordered so that the estimator is invariant to permutation of the variables.

For a sample covariance matrix $\mathbf{S} = (s_{ij})$, the thresholding operator T_λ for an $\lambda \geq 0$ is defined by

$$T_\lambda(\mathbf{S}) = \left[s_{ij}\mathbf{1}(|s_{ij}| \geq \lambda)\right],$$

so that thresholding $\mathbf{S}$ at λ amounts to replacing by zero all entries with absolute value less than λ. Note that T_λ preserves symmetry and is invariant under permutations of variable labels, but does not necessarily preserve positive-definiteness. Its biggest advantage is its simplicity as it carries no major computational burden compared to its competitors like the penalized likelihood with the Lasso penalty (Huang et al., 2006; Rothman et al., 2008).

Just as in banding, one can show the consistency of the threshold estimator in the operator norm provided that $\log p/n \to 0$. The convergence is uniform over the following class of matrices invariant under permutations of indices and satisfying a notion of "approximate sparsity":

$$\mathcal{U}_\tau(q, s_0(p), C) = \left\{\mathbf{\Sigma} : \sigma_{ii} \leq C, \sum_{j=1}^{p} |\sigma_{ij}|^q \leq s_0(p), \text{ for all } \quad i\right\}, \qquad (6.3)$$

for $0 \leq q \leq 1$. Note that for $q = 0$, the inequality in 6.3 reduces to

$$\sum_{j=1}^{p} \mathbf{1}(\sigma_{ij} = 0) \leq s_0(p),$$

which shows that the class is more sparse for smaller values of $s_0(p)$. In this sense and because of dependence of $s_0(p)$ on p, the above condition could be viewed as an implicit definition of *sparsity*. A similar but smaller class of sparse matrices is needed in the study of norm consistency of the thresholded precision matrices:

$$\mathcal{U}_\tau(q, s_0(p), C, \varepsilon) = \{\mathbf{\Sigma} : \mathbf{\Sigma} \in \mathcal{U}_\tau(q, s_0(p), C) \quad \text{and} \quad \lambda_{\min}(\mathbf{\Sigma}) \geq \varepsilon > 0\}.$$

The following consistency result due to Bickel and Levina (2008b) parallels the results for banding stated in Theorem 12.

Theorem 13 *Suppose $\mathbf{Y}$ is Gaussian. Then, uniformly on $\mathcal{U}_\tau(q, s_0(p), C)$, for sufficiently large C', if*

$$\lambda_n = C'\sqrt{\frac{\log p}{n}}, \tag{6.4}$$

and $\frac{\log p}{n} = o(1)$, we have,

$$||T_{\lambda_n}(\mathbf{S}) - \mathbf{\Sigma}|| = O_P\left(s_0(p)\left(\frac{\log p}{n}\right)^{(1-q)/2}\right), \tag{6.5}$$

and uniformly on $\mathcal{U}_\tau(q, s_0(p), C, \varepsilon)$

$$||T_{\lambda_n}(\mathbf{S})^{-1} - \mathbf{\Sigma}^{-1}|| = O_P\left(s_0(p)\left(\frac{\log p}{n}\right)^{(1-q)/2}\right). \tag{6.6}$$

It is interesting to note how the rate of convergence depends explicitly on the dimension p and $s_0(p)$, the number of nonzero elements of the population covariance matrix $\mathbf{\Sigma}$. Furthermore, the rate here is similar to rates of form $\sqrt{\frac{s \log p}{n}}$ obtained for sparse covariance (precision) matrix estimators in several other situations where s is the number of nonzero off-diagonal entries of $\mathbf{\Sigma}$, its inverse, or Cholesky factor (Lam and Fan, 2009).

It is instructive to compare the rates of convergence of the banded and thresholded estimators in Theorems 12 and 13 for the class of covariance matrices:

$$\mathcal{V}(\alpha, C, \varepsilon) = \left\{\mathbf{\Sigma} \in \mathcal{C}(\varepsilon) : |\sigma_{ij}| \leq C|i-j|^{-(\alpha+1)}, \text{for all } i, j \text{ with } |i-j| \geq 1\right\}, \tag{6.7}$$

which can be shown to satisfy both the banding and thresholding conditions so long as $q > (\alpha+1)^{-1}$. Comparing the exponents of the two rates of convergence in these theorems, it can be seen that in situations where the variables are ordered, the banding could do slightly better if

$$1 - q < \frac{\alpha}{\alpha+1}.$$

However, in the genuinely sparse case, that is, when $\alpha \to \infty$ or when q is approximately zero, then both bounds approach $\left(\frac{\log p}{n}\right)^{1/2}$, so that the two methods become comparable.

A broader class of *generalized thresholding operators* that combine thresholding with shrinkage is introduced in Rothman et al. (2009), it covers as special cases, hard- and soft-thresholding, SCAD, and adaptive Lasso. More precisely, for a $\lambda \geq 0$, the generalized thresholding operator is a function $s_\lambda : \mathbf{R} \to \mathbf{R}$ satisfying the following conditions for all $x \in \mathbf{R}$:

(i) $|s_\lambda(x)| \le |x|$,
(ii) $s_\lambda(x) = 0 \quad \text{for} \quad |x| \le \lambda$,
(iii) $|s_\lambda(x) - x| \le \lambda$.

It is evident that the conditions (i)–(iii), respectively, enforce shrinkage, thresholding, and restriction on the amount of shrinkage to no more than λ.

The generalized thresholded covariance estimator is obtained by applying the generalized thresholding operator $s_\lambda(\cdot)$ to each entry of the sample covariance matrix $\mathbf{S}$:

$$s_\lambda(\mathbf{S}) = (s_\lambda(s_{ij})).$$

The generalized thresholding of the sample covariance matrix in high dimensions has good theoretical properties and carries almost no computational burden other than selection of the tuning parameter. In fact, Rothman et al. (2009) show that under conditions (i)–(iii) and those in Theorem 13, the convergence and the rate in (6.5) hold when $T_\lambda(\mathbf{S})$ is replaced by $s_\lambda(\mathbf{S})$, the generalized thresholded covariance estimator. The result shows explicitly the tradeoff among the sparsity of the true covariance matrix $s_0(p)$, dimension p, and the sample size n. Moreover, the generalized thresholding has the *sparsistency* property, in the sense that it estimates the true zero entries as zeros with probability tending to 1.

Theorem 14 *Suppose* $\boldsymbol{Y}$ *is Gaussian with* $\sigma_{ii} \le C$ *for all* i *and* $s_\lambda(\cdot)$ *satisfies the conditions (i)–(iii). If for a sufficiently large* C, $\lambda_n = C\sqrt{\frac{\log p}{n}} = o(1)$, *then*

$$s_{\lambda_n}(s_{ij}) = 0, \quad \textit{for all} \quad (i, j) \quad \textit{such that} \quad \sigma_{ij} = 0,$$

with probability tending to 1.

We note that for sparsistency to hold it is only required that the variances are bounded, and $\boldsymbol{\Sigma}$ does not need to be in the uniformity class.

The choice of the threshold parameter $\lambda = \lambda_n$ is important in implementing this procedure in practice. The problem of threshold selection is hard to deal with analytically. A cross-validation method is proposed by Bickel and Levina (2008b) to select the threshold parameter. They use the Frobenius norm to partly analyze its theoretical and numerical performance. This approach is usually called *universal thresholding* since the same threshold λ_n is used for all entries of the sample covariance matrix. However, the entries of the sample covariance matrix are heteroscedastic in the sense that s_{ij}'s have variances which depend on the distribution of the pairs (Y_i, Y_j) as quantified by

$$\theta_{ij} = \text{var}[(Y_i - \mu_i)(Y_j - \mu_j)]. \tag{6.8}$$

For example, it can be shown that when $\boldsymbol{Y}$ is Gaussian, then

$$\sigma_{ii}\sigma_{jj} \leq \theta_{ij} \leq 2\sigma_{ii}\sigma_{jj},$$

so that the condition $\max_i \sigma_{ii} \leq C$ in $\mathcal{U}_\tau(q, s_0(p), C)$ ensures that the variances of the entries of the sample covariance are uniformly bounded. However, outside this class it makes sense to threshold differentially to account for the possible difference in the variability of the different entries.

Recently, an adaptive thresholded estimator of the form

$$s^*_{\lambda_{ij}}(\mathbf{S}) = (s_{\lambda_{ij}}(s_{ij})), \tag{6.9}$$

where each entry has a different threshold λ_{ij} has been introduced by Cai and Liu (2011). The individual thresholds λ_{ij} are fully data-driven and adapt to the variability of individual entries of the sample covariance. The choice of regularization parameters or the thresholds λ_{ij} in (6.9) are based on an estimator of the variance of the entries θ_{ij} of the sample covariance matrix. More specifically,

$$\lambda_{ij} = \delta\sqrt{\frac{\widehat{\theta}_{ij}\log p}{n}}, \tag{6.10}$$

where δ is a tuning parameter that can be fixed at 2 or chosen empirically through cross-validation, and

$$\widehat{\theta}_{ij} = \frac{1}{n}\sum_{k=1}^{n}[(Y_{ki} - \bar{Y}_i)(Y_{kj} - \bar{Y}_j) - s_{ij}]^2,$$

are estimates of θ_{ij}, where for an $n \times p$ data matrix $\boldsymbol{Y} = (Y_{ki})$, $\bar{Y}_i$ stands for the mean of the entries of the ith column.

It can be shown that the ensuing adaptive estimator (6.9) attains the optimal rate of convergence under the spectral norm for a class of sparse covariance matrices larger than the uniformity class $\mathcal{U}_\tau(q, s_0(p), C)$. The new class defined by

$$\mathcal{U}_\tau{}^*(q, s_0(p)) = \left\{ \mathbf{\Sigma} : \max_i \sum_{j=1}^{p}(\sigma_{ii}\sigma_{jj})^{(1-q)/2}|\sigma_{ij}|^q \leq s_0(p) \right\} \tag{6.11}$$

no longer requires that the variances σ_{ii} to be uniformly bounded and allows $\max_i \sigma_{ii} \to \infty$.

Analogs of Theorems 13 and 14 are proved in Cai and Liu (2011) for adaptive thresholding. They show that $\delta = 2$ is the optimal choice for support recovery in the sense that an adaptive thresholding estimator with any smaller choice of δ would fail to recover the support of $\mathbf{\Sigma}$ exactly with probability converging to 1.

6.4 LOW-RANK PLUS SPARSE COVARIANCE MATRICES

Thus far, sparsity has been the key assumption in regularizing and consistently estimating a high-dimensional covariance matrix. In this section, we deal with a larger class of covariance matrices which can be written as the sum of low-rank and sparse matrices.

The need for this larger class arises in situations where the underlying covariance matrix is not sparse. A prominent example of this occurs in the factor models discussed in Section 3.7 where the covariance matrices are of the form $\boldsymbol{\Sigma} = \Lambda\Lambda' + \Psi$, that is a low-rank matrix plus a diagonal (sparse) matrix. However, when Ψ is sparse instead of being diagonal, the strict factor model does not hold and the more general model is referred to as an *approximate factor model* (Chamberlain and Rothschild, 1983).

The goal in the approximate factor model is to find q, the number of common factors, and consistently estimate the covariance matrix of the data $\boldsymbol{Y}_1, \cdots, \boldsymbol{Y}_n$ when the dimension p could diverge at a rate faster than n. Using the close connection between the (approximate) factor models and the PCA, one can rely on PCs as the estimators of the factor loadings and PC scores as the estimates of the common factors. More precisely, the method proposed by Fan et al. (2013) proceeds as follows:

1. Start with the spectral decomposition of the sample covariance matrix of the data,

$$\mathbf{S} = \sum_{i=1}^{q} \widehat{\lambda}_i \widehat{e}_i \widehat{e}_i' + \widehat{R},$$

where q is the number of selected PCs and $\widehat{R} = (r_{ij})$ is the matrix of residuals.

2. Apply the adaptive thresholding (6.9) to the matrix of residuals to obtain

$$\widehat{R}^{\delta} = (\widehat{r}_{ij}^{\delta}),$$

where

$$\widehat{r}_{ij}^{\delta} = \widehat{r}_{ij} I(|\widehat{r}_{ij}| \geq \delta_{ij}),$$

and $\delta_{ij} > 0$ is an entry-dependent adaptive threshold. Recall that the universal (standard) thresholding corresponds to $\delta_{ij} = \delta$ and a useful example of adaptive thresholding is

$$\delta_{ij} = \delta(\widehat{r}_{ii}\widehat{r}_{jj})^{1/2}$$

for a given $\delta > 0$.

3. An estimator of a low rank plus sparse matrix $\boldsymbol{\Sigma}$ is given by

$$\widehat{\boldsymbol{\Sigma}}^{\delta} = \sum_{i=1}^{q} \widehat{\lambda}_i \widehat{e}_i \widehat{e}_i' + \widehat{R}^{\delta}. \tag{6.12}$$

The estimator is obtained by adaptive thresholding the residual matrix, after taking out the first q PCs. Finding q using data-based methods is an important familiar and well-studied topic in the literature of PCA and factor analysis.

Interestingly, the above procedure is general enough to include some of the most important earlier covariance estimation methods. In fact, for various choices of (δ, q) in (6.12) one obtains the following common covariance estimators:

1. When $\delta = 0$, the estimator reduces to the sample covariance matrix $\mathbf{S}$.
2. When $\delta = 1$, the estimator becomes that based on the standard factor model
3. When $q = 0$, the estimator reduces to the thresholded estimator of Bickel and Levina (2008a) or the adaptive thresholded estimator of Cai and Liu (2011) with a slight and appropriate modification of the thresholding parameter that accounts for the standard errors of the entries of the sample covariance matrix. The discussion paper by Fan et al. (2013) offers the most current results and ongoing research in the area of high-dimensional covariance estimation.

6.5 FURTHER READING

There are several componentwise regularization methods for estimating a sparse or approximately sparse covariance matrix. Thresholding is by far the most popular; in fact, there are several types of thresholding of the sample covariance matrix including hard-thresholding (Bickel and Levina, 2008b; El Karoui, 2008), soft-thresholding with generalizations (Rothman et al., 2009), and adaptive thresholding (Cai and Liu, 2011). In general, threshold estimators have a very low computational cost and good rates of convergence under high-dimensional asymptotics, but may have negative eigenvalues in finite samples. Constructing a sparse covariance estimator that is positive definite in finite samples is desirable when the estimator is used in practice. Rothman (2012) relies on a Lasso-type penalty to encourage sparsity and a logarithmic barrier function to enforce positive-definiteness. Xue et al. (2012) relies on alternating direction method to ensure the positive-definiteness of ℓ_1-penalized covariance estimator.

PROBLEMS

1. For $\boldsymbol{X}_1, \ldots, \boldsymbol{X}_n \sim N_p(\mathbf{0}, \mathbf{\Sigma})$ and $\boldsymbol{S}$ the sample covariance matrix, compute

$$E||B_\ell(\boldsymbol{S}) - \mathbf{\Sigma}||^2,$$

where $0 < \ell \leq p$ is the bandwidth of the banded estimator.

2. Simulate $X_1, \ldots, X_n \sim N_p(\mathbf{0}, \boldsymbol{\Sigma})$ for each of the following cases. Let S be the sample covariance matrix and p_i be the ith percentile of its diagonal entries.

(a) $\boldsymbol{\Sigma}$ is the compound symmetry matrix with $\sigma^2 = 1, \rho = 0.5$, and $(n, p) = (50,25)$. Hard-threshold the sample covariance matrix with the threshold parameter $\lambda = p_1$, compute and count the number of negative eigenvalues of the soft-thresholded covariance matrix. Examine the sparsity of

$$\mathbf{S}^+ = \sum_{i=1}^{p} \max(\hat{\lambda}_i, 0)\hat{e}_i\hat{e}_i'$$

obtained from the spectral decomposition of S replacing the negative eigenvalues by zero.

(b) Repeat part (a) with λ selected via fivefold cross-validation.

(c) Repeat (a) and (b) with $(n, p) = (25, 50)$.

3. Repeat (a)–(c) above when $\boldsymbol{\Sigma}$ is the AR(1) covariance matrix with $\sigma^2 = 1$, $\rho = 0.5$.

4. Repeat Problems 2 and 3 when the hard-thresholding function is replaced by soft-thresholding, SCAD, and adaptive thresholding functions.

CHAPTER 7

MULTIVARIATE REGRESSION: ACCOUNTING FOR CORRELATION

Multivariate regression analysis is concerned with modeling the relationship among several responses and the same set of predictors, say, q responses and p predictors. The multivariate responses measured on the same subject are correlated and one must find ways to exploit directly the dependence in the various stages of inference. The need for vector responses arises in the longitudinal and panel studies and variety of application areas such as chemometrics, financial econometrics, psychometrics, and social sciences where the interest lies in predicting multiple responses with a single set of covariates (Breiman and Friedman, 1997; Johnson and Wichern, 2008; Anderson, 2003).

Prediction in the context of multivariate regression requires estimating at least the pq parameters in the regression coefficient matrix $\mathbf{B}$ which is challenging when there are many predictors and responses. One may start with variable selection methods like the stepwise selection and AIC to reduce the number of regression coefficients. However, due to the discrete nature of such procedures (Breiman, 1996) they are unstable in the sense that small perturbations in the data may lead to very different estimators or predictors.

It is a rather curious fact that once a model is formulated, the standard estimation methods for multivariate regression such as the weighted least-squares (WLS) and maximum likelihood estimators (MLEs), do not exploit the dependence in the responses, so that the resulting estimators are equivalent to regressing separately each response on the p predictors. In contrast, the *reduced rank regression* (RRR) (Anderson, 1951), which finds the least-squares estimators by imposing a rank constraint on the coefficient matrix, does employ implicitly the correlation in the responses. In fact,

High-Dimensional Covariance Estimation, First Edition. Mohsen Pourahmadi.

its solution is related to the low-rank approximation property of the singular value decomposition (SVD) of the regression coefficient matrix, the canonical correlation analysis (CCA) of the multiple responses and predictors and hence to a host of other dimension-reduction methods commonly referred to as linear factor regression in which the response is regressed on a small number of linear combinations of the predictors called the factors.

In this chapter, we review the RRR and various ways to regularize the regression coefficient matrix. Particular attention is paid to using the Lasso penalty on its singular values and singular vectors. We emphasize the importance of penalizing simultaneously the $\boldsymbol{B}$ as well as the $q \times q$ covariance (precision) matrix of the response vector. Two versions of multivariate regression with covariance estimation (MRCE) algorithm (Rothman et al., 2010b) are presented and illustrated using two datasets.

7.1 MULTIVARIATE REGRESSION AND LS ESTIMATORS

In this section, we introduce the multivariate regression model and review some of its basic properties.

We start by introducing some notation. Let $x_i = (x_{i1}, \ldots, x_{ip})'$ denote the vector of predictors, $\boldsymbol{Y}_i = (Y_{i1}, \ldots, Y_{iq})'$ be the vector of responses. The multivariate linear regression model relates the response $\boldsymbol{Y}_i$ to the predictors x_i via

$$\boldsymbol{Y}_i = \mathbf{B}'x_i + \boldsymbol{\varepsilon}_i, \quad \text{for i} = 1, \ldots, n,$$

where $\mathbf{B}$ is a $p \times q$ regression coefficient matrix and $\boldsymbol{\varepsilon}_i = (\varepsilon_1, \ldots, \varepsilon_q)'$ is the vector of errors for the ith subject. As usual, it is more convenient to write the model in matrix notation. Let $\boldsymbol{Y}$ denote the $n \times q$ response matrix where its ith row is $\boldsymbol{Y}_i'$, $\boldsymbol{X}$ be the $n \times p$ predictor matrix where its ith row is x_i', and let $\mathbf{E}$ denote the $n \times q$ random error matrix where its ith row is $\boldsymbol{\varepsilon}_i'$, then the multivariate regression model in matrix form is given by

$$\boldsymbol{Y} = \boldsymbol{X}\mathbf{B} + \mathbf{E}. \tag{7.1}$$

Note that when $q = 1$, it reduces to the multiple linear regression model where $\mathbf{B}$ is a p-dimensional regression coefficient vector.

For simplicity, in this chapter we assume that $\boldsymbol{X}$ has full-rank, the columns of $\boldsymbol{X}$ and $\boldsymbol{Y}$ have been centered and hence the intercept terms are omitted.

In what follows, we make the standard assumption that $\boldsymbol{\varepsilon}_1, \ldots, \boldsymbol{\varepsilon}_n$ are $i.i.d \sim N_q(\mathbf{0}, \boldsymbol{\Sigma})$, so that the covariance matrix of the response vector for each subject is $\boldsymbol{\Sigma} = (\sigma_{ij})$. The fact that the q responses from the ith sample are correlated suggests that one should not be estimating columns of $\mathbf{B}$ via q separate multiple regressions. Evidently, this amounts to ignoring the correlation and such an estimator might be inferior to jointly estimating all columns of $\mathbf{B}$ and by accounting for the dependence in the responses in some manners. Surprisingly, this is exactly what happens so that the multivariate aspect of the response vector does not help with improving in the estimation $\boldsymbol{B}$ or the prediction of $\boldsymbol{Y}$.

To explain this phenomenon more clearly, note that the kth column of the matrix $\mathbf{B}$, denoted by $\mathbf{B}_{\cdot k}$, is the vector regression coefficients in

$$\boldsymbol{Y}_{\cdot k} = \boldsymbol{X}\mathbf{B}_{\cdot k} + \mathbf{E}_{\cdot k}, k = 1, \ldots, q, \operatorname{cov}(\mathbf{E}_{\cdot k}) = \sigma_{kk} I_n$$

when the kth response is regressed on all the predictors. From the standard theory, the least-squares estimator of $\mathbf{B}_{\cdot k}$ is

$$\widehat{\mathbf{B}}_{\cdot k} = (\boldsymbol{X}'\boldsymbol{X})^{-1}\boldsymbol{X}'\boldsymbol{Y}, k = 1 \ldots, q.$$

Collecting these individual least-squares estimators in a matrix, we arrive at an estimate of $\mathbf{B}$ as

$$\widehat{\mathbf{B}} = (\widehat{\mathbf{B}}_{\cdot 1}, \ldots, \widehat{\mathbf{B}}_{\cdot q}),$$

which actually has the important property of minimizing the criterion

$$Q(\boldsymbol{B}) = \operatorname{tr}(\boldsymbol{Y} - \boldsymbol{X}\mathbf{B})'(\boldsymbol{Y} - \boldsymbol{X}\mathbf{B}) = \sum_{1}^{q}(\boldsymbol{Y}_{\cdot k} - \boldsymbol{X}\mathbf{B}_{\cdot k})'(\boldsymbol{Y}_{\cdot k} - \boldsymbol{X}\mathbf{B}_{\cdot k}).$$

Evidently, using $\widehat{\mathbf{B}}$, the predicted values of $\boldsymbol{Y}$ given by $\widehat{\boldsymbol{Y}}_{\mathrm{OLS}} = \boldsymbol{X}\widehat{\mathbf{B}}$, does not take advantage of the correlation in $\boldsymbol{Y}$. As a remedy, the Curds and Whey (C&W) method proposed by Breiman and Friedman (1997) predicts a multivariate response vector using $\tilde{\boldsymbol{Y}} = \widehat{\boldsymbol{Y}}_{\mathrm{OLS}}\mathbf{M}$ where $\widehat{\boldsymbol{Y}}_{\mathrm{OLS}}$ is the ordinary least-square predictor and $\mathbf{M}$ is a $q \times q$ shrinkage matrix estimated from the data through the canonical correlations in (7.9).

One might think that estimating the coefficient matrix $\mathbf{B}$ by means of the WLS or the maximum likelihood estimation (Anderson, 2003) could exploit better the dependence or the multivariate character of the response. Thus, minimizing the WLS criterion:

$$Q_{\boldsymbol{\Sigma}}(\mathbf{B}) = \operatorname{tr}\left[(\boldsymbol{Y} - \boldsymbol{X}\mathbf{B})'\boldsymbol{\Sigma}^{-1}(\boldsymbol{Y} - \boldsymbol{X}\mathbf{B})\right], \tag{7.2}$$

it follows that the WLS estimator $\widehat{\mathbf{B}}_{\mathrm{WLS}}$ satisfies (Problem 1) the normal equations:

$$(\boldsymbol{X}'\boldsymbol{X})\widehat{\mathbf{B}}_{\mathrm{WLS}} = \boldsymbol{X}'\boldsymbol{Y}. \tag{7.3}$$

Recall that the normal log-likelihood function of $(\mathbf{B}, \boldsymbol{\Sigma})$ is

$$\ell(\mathbf{B}, \boldsymbol{\Sigma}) = -\operatorname{tr}\left[\frac{1}{n}(\boldsymbol{Y} - \boldsymbol{X}\mathbf{B})'\boldsymbol{\Sigma}(\boldsymbol{Y} - \boldsymbol{X}\mathbf{B})\right] + \log|\boldsymbol{\Sigma}|. \tag{7.4}$$

Comparing (7.2) and (7.4), it follows that the MLE $\widehat{\mathbf{B}}_{\text{MLE}}$ and the WLS estimator are precisely the same as the ordinary least-squares estimator:

$$\widehat{\boldsymbol{B}}_{\text{MLE}} = \widehat{\mathbf{B}}_{\text{OLS}} = (\boldsymbol{X}'\boldsymbol{X})^{-1}\boldsymbol{X}'\boldsymbol{Y}.$$

Again, this amounts to performing separate ordinary least-squares estimates for each of the q response variables. In short, the MLE of $\mathbf{B}$ does not take advantage of the dependence structure of the responses. The MLE of $\boldsymbol{\Sigma}$ is given by

$$\widehat{\boldsymbol{\Sigma}} = \frac{1}{n}(\boldsymbol{Y} - \boldsymbol{X}\widehat{\mathbf{B}}_{\text{OLS}})'(\boldsymbol{Y} - \boldsymbol{X}\widehat{\mathbf{B}}_{\text{OLS}}). \tag{7.5}$$

For high-dimensional data, particularly when p and q are larger than n, the matrix $\mathbf{B}$ cannot be estimated uniquely since $\boldsymbol{X}$ is not of full column rank, and $\mathbf{S}$ the sample covariance matrix is not invertible. In these situations, the traditional estimators for $\mathbf{B}$ and $\boldsymbol{\Sigma}$ with pq and $q(q+1)/2$ parameters, respectively, have rather poor computational and statistical performances and they are not suitable for prediction and inference. Thus, one must seek suitable alternative estimators based on the ideas of *shrinkage and regularization*. Historically, this has been done by regularizing separately either $\mathbf{B}$ or $\boldsymbol{\Omega}$, but not both of them simultaneously.

7.2 REDUCED RANK REGRESSIONS (RRR)

In this section, the technique of RRR and its close connections with SVD, PCA, factor analysis, CCA, and other dimension-reduction methods are discussed.

Most recent methods of reducing the number of parameters in $\mathbf{B}$ are inspired by the idea of RRR (Anderson, 1951; Izenman, 1975; Reinsel and Velu, 1998). The RRR amounts to estimating $\mathbf{B}$ by solving a constrained least-squares problem, that is, minimizing $\operatorname{tr}\left[(\boldsymbol{Y} - \boldsymbol{X}\mathbf{B})'(\boldsymbol{Y} - \boldsymbol{X}\mathbf{B})\right]$ over $\mathbf{B}$ subject to a rank constraint

$$\operatorname{rank}(\mathbf{B}) = r, \quad \text{for some} \quad r \leq m = \min(p, q).$$

The solution, denoted by $\widehat{\mathbf{B}}_{(r)}$, is known (Reinsel and Velu, 1998) to be closely related to the lease squares estimates:

$$\widehat{\mathbf{B}}_{(r)} = (\boldsymbol{X}'\boldsymbol{X})^{-1}\boldsymbol{X}'\boldsymbol{Y}\boldsymbol{H}\boldsymbol{H}' = \widehat{\boldsymbol{B}}_{\text{OLS}}\boldsymbol{H}\boldsymbol{H}', \tag{7.6}$$

where $\boldsymbol{H} = (\boldsymbol{h}_1, \ldots, \boldsymbol{h}_r)$ and $\boldsymbol{h}_k$ is the normalized eigenvector corresponding to the kth largest eigenvalue of the matrix $\boldsymbol{Y}'\boldsymbol{X}(\boldsymbol{X}'\boldsymbol{X})^{-1}\boldsymbol{X}'\boldsymbol{Y}$.

Another useful representation of $\widehat{\mathbf{B}}_{(r)}$ exploits the SVD of the rank m matrix (Aldrin, 2000)

$$(\boldsymbol{X}'\boldsymbol{X})^{1/2}\widehat{\mathbf{B}}_{\text{OLS}} = \sum_{1}^{m} \lambda_k \boldsymbol{u}_k \mathbf{v}_k'.$$

Pre-multiplying both sides by $(X'X)^{-1/2}$ it follows that $\widehat{\mathbf{B}}_{\text{OLS}}$ can be decomposed into a sum of m rank-one matrices as

$$\widehat{\mathbf{B}}_{\text{OLS}} = \sum_{k=1}^{m} \widehat{\mathbf{B}}_k,$$

where $\widehat{\mathbf{B}}_k = \lambda_k (X'X)^{-1/2} \boldsymbol{u}_k \mathbf{v}_k'$. Then, the rank r least-squares estimate of $\boldsymbol{B}$ is given by

$$\widehat{\mathbf{B}}_{(r)} = \sum_{k=1}^{r} \widehat{B}_k, \tag{7.7}$$

which amounts to *hard-thresholding* the SVD of the least-squares estimator. Note that for $r = m$, the RRR estimate reduces to the OLS, and $r = 0$ corresponds to a pure intercept model with $\widehat{\mathbf{B}}_{(r)} = 0$. The rank r in (7.7) has a clear role as a tuning parameter balancing between the bias and variance of prediction. As a compromise, a *soft-thresholding* of the least-squares estimators can be introduced by giving nonzero weights to the m components to obtain

$$\widehat{\mathbf{B}}_S = \sum_{k=1}^{m} w_k \widehat{\mathbf{B}}_k, \tag{7.8}$$

where w_k is the weight assigned to the kth component. A method for estimating the weights as regression coefficients using PCA of $\boldsymbol{X}$ is given in Aldrin (2000).

The idea of RRR is also connected to the CCA between the q responses $\boldsymbol{Y}$ and the p covariates $\boldsymbol{X}$. As a data-reduction technique introduced by Hotelling (1935), CCA has the goal of finding a sequence of uncorrelated linear combinations of the predictors and the responses, namely $\boldsymbol{X}\mathbf{v}_k, \boldsymbol{Y}\boldsymbol{u}_k, k = 1, \ldots, m = \min(p, q)$, such that their squared correlations

$$R_k^2 = \text{corr}^2(\boldsymbol{Y}\boldsymbol{u}_k, \boldsymbol{X}\mathbf{v}_k) \tag{7.9}$$

are maximized. The solutions $(\boldsymbol{u}_k, \mathbf{v}_k, R_k^2), k = 1, \ldots, m$ are found using a generalized SVD of the cross-covariance matrix $\boldsymbol{Y}'\boldsymbol{X}/n$. However, the interpretation of these canonical variates or linear combinations of the responses and covariates is a challenging task just as in linear regression and PCA when both p, q are large.

The RRR and CCA are specific examples of a broad family of dimension-reduction methods known as *linear factor regression* in which the response $\boldsymbol{Y}$ is regressed on a small number of linearly transformed predictors $\boldsymbol{X}$ referred to as the factors. They involve rewriting the multivariate regression model as

$$\boldsymbol{Y} = \boldsymbol{X}\Gamma'\Gamma\mathbf{B} + \mathbf{E} = \mathbf{F}\tilde{\mathbf{B}} + \mathbf{E}, \tag{7.10}$$

where Γ is a $p \times r$ matrix with orthonormal columns and $r \leq m$. The columns of $\mathbf{F} = \boldsymbol{X}\Gamma'$ can be interpreted as the r latent factors that explain the variation in the responses and $\tilde{\mathbf{B}}$ is an $r \times q$ matrix of factor loadings or transformed regression coefficients. It is expected that the correlations between the q responses are also taken into account by these common latent factors. A two-step procedure is used usually to estimate the linear factor regression models where first a Γ and hence the factors are determined in some manners (such as using the PCs) and then the $\tilde{\mathbf{B}}$ is estimated using the least-squares method applied to (7.10) with $\mathbf{F}$ treated as the known design matrix (see Section 3.7).

A recent method due to Yuan et al. (2007), called *factor estimation and selection*, avoids the two-stage estimation procedure of the previous section by regressing the response Y against a small number of linearly transformed predictors. It is based on the SVD representation

$$\mathbf{B} = \mathbf{UDV}' = \sum_{k=1}^{m} d_k \boldsymbol{u}_k \mathbf{v}_k',$$

(see Section 4.3), which reveals a powerful factor model interpretation of the multivariate regression in the sense that the coefficient matrix is decomposed into m orthogonal layers of decreasing importance determined by the relative size of d_k's. In this setup, using the SVD of $\mathbf{B}$ and post-multiplying the multivariate regression model (7.1) by $\mathbf{V}$, it follows that

$$\boldsymbol{Y}\mathbf{V} = \boldsymbol{X}\mathbf{UD} + \mathbf{EV}, \tag{7.11}$$

which suggests that for each layer k, the elements in $\boldsymbol{u}_k$ can be viewed as the predictor effects, the elements in $\mathbf{v}_k$ as the response effects, and the singular value d_k indicates the relative importance of the kth layer.

7.3 REGULARIZED ESTIMATION OF B

The idea of penalized linear regression is even more appealing in the multivariate regression setup since there are more meaningful ways to introduce penalty functions on the matrix of coefficients. For example, one may impose various Lasso penalties directly on $\mathbf{B}$ or its SVD components. The latter works only for multivariate regression.

Regularizing $\mathbf{B}$ directly, without resorting to its *rank* or SVD, amounts to solving the following constraint optimization problem:

$$\underset{\mathbf{B}}{\operatorname{argmin}} \left\{ \operatorname{tr}\left[(\boldsymbol{Y} - \boldsymbol{X}\mathbf{B})'(\boldsymbol{Y} - \boldsymbol{X}\mathbf{B})\right] + \lambda C(\mathbf{B}) \right\}, \tag{7.12}$$

with $C(\mathbf{B})$ a scalar function and λ is a nonnegative penalty.

The first natural and familiar constraint on the coefficient matrix is $C(\mathbf{B}) = \sum_{j,k} b_{jk}^2$ so that (7.12) leads to the closed-form solution of the ridge regression problem. In analogy with multiple regression, it opens up the possibility of using other ℓ_p

norms of **B** as penalty functions. For example, the ℓ_1 norm $C(\mathbf{B}) = \sum_{j,k} |b_{jk}|$ leads to the Lasso estimate:

$$\widehat{\mathbf{B}} = \underset{\mathbf{B}}{\operatorname{argmin}} \left\{ \operatorname{tr}\left[(\boldsymbol{Y} - \boldsymbol{X}\mathbf{B})'(\boldsymbol{Y} - \boldsymbol{X}\mathbf{B})\right] + \lambda \sum_{j,k} |b_{jk}| \right\}. \tag{7.13}$$

If needed, one may assign different weights to different parameters and use the adaptive Lasso (Zou, 2006) which amounts to setting $C(\mathbf{B}) = \sum_{j,k} w_{jk}|b_{jk}|$, where w'_{jk}, s are chosen adaptively using the data.

Other forms of the constraint function which seem to make a compromise between the Lasso and the ridge regression are

- The *bridge regression* takes $C(\mathbf{B}) = \sum_{j,k} |b_{jk}|^{\gamma}$ where $1 \le \gamma \le 2$; the elastic-net with $C(\mathbf{B}) = \alpha \sum_{j,k} |b_{jk}| + \frac{(1-\alpha)}{2} \sum_{j,k} b_{jk}^2$ for $\alpha \in [0, 1]$ (Zou and Hastie, 2005).
- *Group-wise penalty functions* are particularly suitable for regularizing the multivariate regression parameters. The first example of its kind is the grouped Lasso with $C(\mathbf{B}) = \sum_{j=1}^{p} (b_{j1}^2 + \cdots + b_{jq}^2)^{0.5}$ (Yuan and Lin, 2007). One could also combine the ℓ_1 and ℓ_2 penalties to form the constraint $C(\mathbf{B}) = \alpha C_1(\mathbf{B}) + (1 - \alpha)C_2(\mathbf{B})$ for $\alpha \in [0, 1]$ where $C_1(\mathbf{B}) = \sum_{j,k} |b_{jk}|$ and $C_2(\mathbf{B}) = \sum_{j=1}^{p} (b_{j1}^2 + \cdots + b_{jq}^2)^{0.5}$. Note that the first constraint controls the overall sparsity of the coefficient matrix **B** and the second imposes a group-wise penalty on the rows of **B** which could introduce zeros in some rows of **B** showing that some predictors are irrelevant for all the responses (Peng et al., 2010).

To regularize the components of the SVD of **B**, it is helpful to recall that the RRR finds its rank r least-squares estimate, and the rank of **B** is the number of nonzero singular values or the ℓ_0 norm of the vector of singular values. Motivated by this, one may consider imposing a Lasso penalty on the singular values:

$$C(\mathbf{B}) = \sum_{k=1}^{m} d_k(\mathbf{B}), \tag{7.14}$$

where $d_k(\mathbf{B})$ is the kth singular value of the coefficient matrix **B** (Yuan et al., 2007). It encourages sparsity among the singular values and at the same time shrinks the coefficient estimates, so that it does dimension reduction and estimation simultaneously.

In many practical situations, each layer of the SVD of the matrix **B** provides a distinct channel or pathway of association between the responses and predictors that may involve only a subset of them. Thus, to achieve further dimension reduction and to facilitate interpretation it is desirable to have sparse left and right singular vectors as in Section 4.5 or in Chen et al. (2012). More precisely, one may estimate **B** by minimizing the objective function

$$\frac{1}{2}||\boldsymbol{Y} - \boldsymbol{X}\sum_{k=1}^{r} d_k \boldsymbol{u}_k \mathbf{v}_k'||_F^2 + \sum_{k=1}^{r} \lambda_k P(d_k, \boldsymbol{u}_k, \mathbf{v}_k), \tag{7.15}$$

with respect to the triplet $(d_k, \boldsymbol{u}_k, \mathbf{v}_k)$ for $k = 1, \ldots, r$ where $||\boldsymbol{u}_k||_2 = ||\mathbf{v}_k||_2 = 1$, $P(\cdot)$ is a penalty function, and the λ_k's are the tuning parameters. The optimization problem (7.15) is first solved for the unit rank case, that is, when $r = 1$, and then sequentially the procedure is applied to the residual matrix as in Section 4.5. However, the presence of the matrix $\boldsymbol{X}$ in problem (7.15) could make the optimization problem different from that in Section 4.5. Fortunately, for the Lasso and adaptive Lasso penalty functions, which are proportional to the ℓ_1 norm of an SVD layer $d_k \boldsymbol{u}_k \mathbf{v}_k'$, the matrix $\boldsymbol{X}$ can be absorbed into the singular vectors and hence the optimization problem is reduced to certain Lasso regression problems with explicit solutions as shown in Chen et al. (2012).

Most of the constraints mentioned so far impose penalty only on the regression coefficient matrix **B** without accounting directly for the covariance structure of the q responses. In other words, they mostly ignore the $q(q+1)/2$ parameters in $\boldsymbol{\Sigma}$ whose estimation is a problem of great interest in statistics (Pourahmadi, 2011). The question of how to take advantage of correlations between the response variables to improve predictive accuracy is important in applications, but there are only a few papers devoted to this important subject. Warton (2008) relies on the idea of ridge regression, and Witten and Tibshirani (2009) present a procedure assuming the multivariate normality of the $(p+q)$-dimensional random vector of responses and the predictors and treat the problem as the regularization of the inverse covariance of the joint distribution.

7.4 JOINT REGULARIZATION OF (B, Ω)

In this section, a sparse estimator for **B** is proposed that directly accounts for and penalizes the correlation in the multivariate response in a systematic manner. We follow Rothman et al.'s (2010b) MRCE method which constructs sparse estimates for both matrices by adding separate Lasso penalties to the negative normal log-likelihood.

The objective function to be minimized is proportional to

$$\begin{aligned}\ell(\mathbf{B}, \boldsymbol{\Omega}) = \operatorname{tr}\left[\frac{1}{n}(\boldsymbol{Y} - \boldsymbol{X}\mathbf{B})\boldsymbol{\Omega}(\boldsymbol{Y} - \boldsymbol{X}\mathbf{B})'\right] - \log|\boldsymbol{\Omega}| \\ + \lambda_1 \sum_{j' \neq j} |\omega_{j'j}| + \lambda_2 \sum_{j,k} |b_{jk}|,\end{aligned} \tag{7.16}$$

where $\mathbf{\Omega} = (\omega_{jj'}) = \mathbf{\Sigma}^{-1}$ and λ_1, λ_2 are two tuning parameters to be determined from the data. The Lasso penalty on $\mathbf{B}$ introduces sparsity in the regression coefficients matrix, improves the prediction performance, and possibly facilitates interpretation of the fitted model. Note that the Lasso penalty on the precision matrix involves only its off-diagonal entries, and the optimal solution for $\mathbf{B}$ is a function of $\mathbf{\Omega}$. The solution to problem (7.16) gives a sparse $\mathbf{B}$ that accounts for the correlation structure of the response variables.

The optimization problem in (7.16) is not convex in $(\mathbf{B}, \mathbf{\Omega})$, it is biconvex in the sense that solving it for either $\mathbf{B}$ or $\mathbf{\Omega}$ with the other kept fixed is a convex optimization problem. More precisely, solving problem (7.16) for $\mathbf{\Omega}$ with $\mathbf{B}$ fixed at a chosen point $\mathbf{B}_0$ reduces to the optimization problem:

$$\widehat{\mathbf{\Omega}}(\mathbf{B}_0) = \underset{\mathbf{\Omega}}{\operatorname{argmin}} \left\{ \operatorname{tr}\left(\widehat{\mathbf{\Sigma}}\mathbf{\Omega}\right) - \log|\mathbf{\Omega}| + \lambda_1 \sum_{i \neq j} |\omega_{ij}| \right\}, \tag{7.17}$$

where $\widehat{\mathbf{\Sigma}} = \frac{1}{n}(\boldsymbol{Y} - \boldsymbol{X}\mathbf{B}_0)'(\boldsymbol{Y} - \boldsymbol{X}\mathbf{B}_0)$. This is the familiar ℓ_1-penalized covariance estimation problem discussed in Chapter 5 and has been studied by d'Aspremont et al. (2008), Yuan and Lin (2007), Rothman et al. (2008), and Friedman et al. (2008). The graphical Lasso (Glasso) algorithm of Friedman et al. (2008) will be used to solve (7.17), it is the fastest and the most commonly used algorithm for this problem.

Solving (7.16) for $\mathbf{B}$ with $\mathbf{\Omega}$ fixed at a nonnegative definite $\mathbf{\Omega}_0$ leads to the convex optimization problem:

$$\widehat{\mathbf{B}}(\mathbf{\Omega}_0) = \underset{\mathbf{B}}{\operatorname{argmin}} \left\{ \operatorname{tr}\left[\frac{1}{n}(\boldsymbol{Y} - \boldsymbol{X}\mathbf{B})\mathbf{\Omega}_0(\boldsymbol{Y} - \boldsymbol{X}\mathbf{B})'\right] + \lambda_2 \sum_{j=1}^{p}\sum_{k=1}^{q} |b_{jk}| \right\}. \tag{7.18}$$

The convexity can be shown using the fact that the trace term in the objective function has the Hessian $2n^{-1}\mathbf{\Omega}_0 \otimes \boldsymbol{X}'\boldsymbol{X}$ which is nonnegative definite as the Kronecker product of two symmetric nonnegative definite matrices. A solution is given below by reducing it to a Lasso regression problem.

Define the vector β of length pq by $(b_{11}, b_{21}, \ldots, b_{p1}, \ldots, b_{1q}, b_{2q}, \ldots, b_{pq})'$ and set $\boldsymbol{X}_i = I_{q\times q} \otimes x_i'$, where $\otimes$ is the Kronecker product and $I_{q\times q}$ is the identity matrix. Consider the Cholesky decomposition $\mathbf{\Omega}_0 = L'L$, where L is a $q \times q$ upper triangular matrix. Let $\tilde{\boldsymbol{Y}} = \frac{1}{\sqrt{n}}(\boldsymbol{Y}_1'L', \boldsymbol{Y}_2'L', \ldots, \boldsymbol{Y}_n'L')'$ which is of length qn and $\tilde{\boldsymbol{X}} = \frac{1}{\sqrt{n}}(\boldsymbol{X}_1'L', \ldots, \boldsymbol{X}_n'L')'$. Then, (7.18) can be rewritten more compactly as

$$\widehat{\mathbf{B}}(\mathbf{\Omega}_0) = \underset{\beta}{\operatorname{argmin}} \left\{ \|\tilde{\boldsymbol{Y}} - \tilde{\boldsymbol{X}}\beta\|^2 + \lambda_2 \sum_{j=1}^{pq} |\beta_j| \right\}, \tag{7.19}$$

which is exactly a Lasso regression problem. A simple algorithm for solving such a problem is the coordinate descent algorithm introduced in Chapter 2.

The computational details for solving (7.16) is summarized in the MRCE (Rothman et al., 2010b) algorithm below. We use the ℓ_1 norm of the ridge estimator of $\mathbf{B}$, that is, $\sum_{j,k} |\widehat{b}_{jk}^R|$, to scale the test for convergence in the MRCE algorithm, where $\widehat{\mathbf{B}}_R = (X'X + \lambda_2 \mathbf{I})^{-1} X'Y$, and ϵ is the tolerance parameter set at 10^{-4} by default.

MRCE Algorithm:
With λ_1 and λ_2 fixed, initialize $\widehat{\mathbf{B}}^{(0)} = \mathbf{0}$ and $\widehat{\mathbf{\Omega}}^{(0)} = \widehat{\mathbf{\Omega}}(\widehat{\mathbf{B}}^{(0)})$.

Step 1: Compute $\widehat{\mathbf{B}}^{(m+1)} = \widehat{\mathbf{B}}(\widehat{\mathbf{\Omega}}^{(m)})$ by solving (7.18) using the coordinate descent algorithm.

Step 2: Compute $\widehat{\mathbf{\Omega}}^{(m+1)} = \widehat{\mathbf{\Omega}}(\widehat{\mathbf{B}}^{(m+1)})$ by solving (7.17) using the Glasso algorithm.

Step 3: If $\sum_{i,j} |\widehat{b}_{ij}^{(m+1)} - \widehat{b}_{ij}^{(m)}| < \epsilon \sum_{i,j} \widehat{b}_{i,j}^R$, then stop. Otherwise go to Step 1.

This algorithm uses block-wise coordinate descent to compute a local solution for (7.16) by alternating between Steps 1 and 2. Each iteration will decrease the objective function. In practice, for certain values of the tuning parameters (λ_1, λ_2) the algorithm could be slow and many iterations might be needed for it to converge for high-dimensional data. For such cases, a faster and approximate approach to solving (7.16) is to use a better starting value for $\mathbf{B}$ computed from fitting separate Lasso regressions.

Approximate MRCE:
For fixed values of λ_1 and λ_2,

Step 1: Perform q separate Lasso regressions each with the same optimal tuning parameter $\widehat{\lambda}_0$ selected with a cross-validation procedure. Let $\widehat{\mathbf{B}}_{\widehat{\lambda}_0}$ denote the solution.

Step 2: Compute $\widehat{\mathbf{\Omega}} = \widehat{\mathbf{\Omega}}(\widehat{\mathbf{B}}_{\widehat{\lambda}_0})$ by solving (7.17) using the Glasso algorithm.

Step 3: Compute $\widehat{\mathbf{B}} = \widehat{\mathbf{B}}(\widehat{\mathbf{\Omega}})$ by solving (7.18) using the coordinate descent algorithm.

Unlike the MRCE algorithm , the approximate MRCE algorithm does not alternate among the three steps, but it is iterative only inside the steps. It starts by finding a better optimally tuned Lasso solution $\widehat{\mathbf{B}}_{\widehat{\lambda}_0}$, then proceeds to estimate $\mathbf{\Omega}$ using the Glasso algorithm in Step 2, and finally solves for $\mathbf{B}$ in Step 3. Selecting the two tuning parameters (λ_1, λ_2) needed in the last two steps is discussed next.

The tuning parameters λ_1 and λ_2 are selected using K-fold cross-validation by dividing the data into K groups of nearly equal size. For the kth validation group, the validation prediction error is accumulated over all q responses. Specifically, the optimal tuning parameters $\widehat{\lambda}_1$ and $\widehat{\lambda}_2$ are selected using

$$(\widehat{\lambda}_1, \widehat{\lambda}_2) = \operatorname{argmin}_{\lambda_1,\lambda_2} \sum_{k=1}^{K} \|\boldsymbol{Y}^{(k)} - \boldsymbol{X}^{(k)}\mathbf{B}^{(-k)}_{\lambda_1,\lambda_2}\|_F^2, \tag{7.20}$$

where $\boldsymbol{Y}^{(k)}$ and $\boldsymbol{Y}^{(k)}$ are the matrices of responses and predictors in the kth fold, and $\mathbf{B}^{(-k)}_{\lambda_1,\lambda_2}$ is the estimated regression coefficient matrix computed with observations outside the kth fold, with tuning parameters λ_1 and λ_2. Simulation results suggest that λ_2, which controls the penalty on the regression coefficient matrix, has greater influence on prediction performance than λ_1 (Rothman et al., 2010b). The performance of the fitted model is measured in terms of the model error as in Yuan and Lin (2007) and Rothman et al. (2010b):

$$ME(\widehat{\mathbf{B}}) = \operatorname{tr}\left\{(\widehat{\mathbf{B}} - \mathbf{B})'\boldsymbol{\Sigma}_x(\widehat{\mathbf{B}} - \mathbf{B})\right\}, \tag{7.21}$$

where $\boldsymbol{\Sigma}_x = \boldsymbol{X}'\boldsymbol{X}$. Alternatively, the sparsity recognition performance of $\widehat{\mathbf{B}}$ is measured by the true positive rate (TPR) as well as the true negative rate (TNR) defined, respectively, as

$$\text{TPR}(\widehat{\mathbf{B}}, \mathbf{B}) = \frac{\#\{(i,j) : \widehat{b}_{ij} \neq 0 \quad \text{and} \quad b_{ij} \neq 0\}}{\#\{(i,j) : b_{ij} \neq 0\}}, \tag{7.22}$$

$$\text{TNR}(\widehat{\mathbf{B}}, \mathbf{B}) = \frac{\#\{(i,j) : \widehat{b}_{ij} = 0 \quad \text{and} \quad b_{ij} = 0\}}{\#\{(i,j) : b_{ij} = 0\}}. \tag{7.23}$$

The TPR is the proportion of nonzero elements in $\mathbf{B}$ that $\widehat{\mathbf{B}}$ identifies correctly, while the TNR measures the proportion of zero elements recognized correctly. Of course, it is prudent to consider both of them simultaneously since the trivial estimator $\widehat{\mathbf{B}} = 0$ always has perfect TNR and the OLS estimate always has perfect TPR.

7.5 IMPLEMENTING MRCE: DATA EXAMPLES

In this section, we illustrate the MRCE algorithms using two real datasets.

7.5.1 Intraday Electricity Prices

Accurate modeling and forecasting of the electricity wholesale prices is very important to market participants for the purpose of market trading, investment decision-making, and risk management. As the first example of the use of multivariate regression, we consider the hourly average electricity spot prices collected in the Australian state of New South Wales (NSW) from July 2, 2003 to June 30, 2006, starting at 04:00 A.M. and ending at 03:00 A.M. each day. The dataset consists of 26,352 observations during a period of $T = 1098$ days. (Panagiotelis and Smith, 2008).

Unlike other commodity prices, most electricity spot prices exhibit trend, strong periodicity, intra-day and inter-day serial correlations, heavy tails, skewness, and so on (see Panagiotelis and Smith, 2008, for some empirical evidence). We consider the vector of the log spot prices at hourly intervals during a day as the response vector with $q = 24$. The covariates which may have effects on the spot prices include a simple linear trend, dummy variables for day types (in total 13 dummy variables, representing the seven days of the week and some idiosyncratic public holidays), and eight seasonal polynomials (high-order Fourier terms) for a smooth seasonal effect.

A multivariate regression model is fitted to the log electricity prices by regressing the vector of hourly observations during a day on the same covariates:

$$\boldsymbol{Y}_i = \mathbf{B}' x_i + \boldsymbol{\varepsilon}_i \quad 1 \le i \le T \tag{7.24}$$

where $\boldsymbol{Y}_i$ is a 24×1 vector of log electricity prices on day i and x_i is the corresponding vector of the covariates. We assume $\boldsymbol{\varepsilon} \sim N_q(0, \boldsymbol{\Sigma})$ and use the MRCE algorithms to fit models to this data. We assess the predictive performance of the fitted model via the mean squared prediction error, where the observations from the last 100 days are retained as the test set, while the rest of the observed spot prices are used to estimate the parameters. The tuning parameters are selected from the set $\Lambda = \{2^{-10+20(x-1)/39} : x = 1, 2, \ldots, 40\}$ via 10-fold cross-validation.

The average squared prediction errors at different times of a day based on the observations in the last 100 days are plotted in Figure 7.1 where the results using the

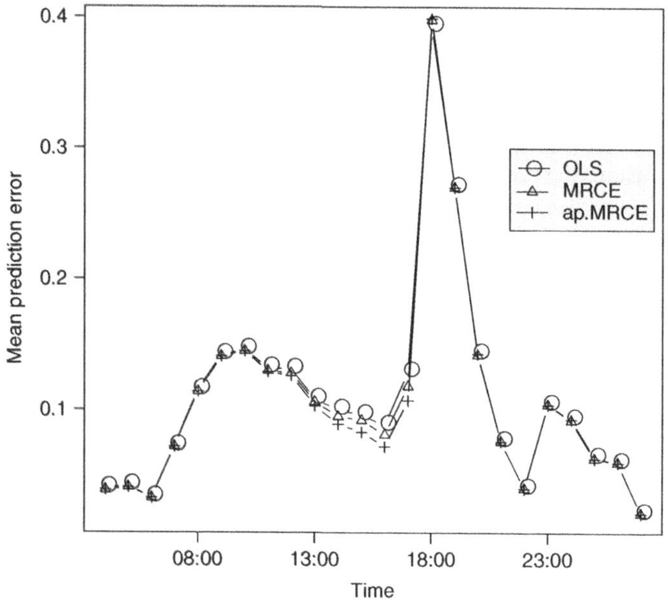

FIGURE 7.1 The average squared prediction error for each hour on a day based on 100 points.

TABLE 7.1 Proportions of zeros in the estimate of the parameters

	MRCE	Approximate MRCE
B	74/528	70/528
$\mathbf{\Omega}$	484/576	506/576

OLS method are also included for comparison. We see that the three methods have similar performances, but compared to the OLS method the two MRCE algorithms slightly improve the prediction accuracy. The overall prediction errors for the OLS, MRCE, and approximate MRCE are 0.110, 0.108, 0.107, respectively, implying that the approximate MRCE method does the best in this example. The proportions of zeros in the estimated regression coefficient matrix as well as the regularized precision matrix are presented in Table 7.1. We see that the estimated coefficient matrices for both MRCE methods are fairly sparse, implying that some of the covariates do have impact on the hourly spot price. However, the estimated precision matrices are very sparse with the positions of nonzero entries displayed in Figure 7.2. We note that both MRCE algorithms produce block diagonal structures for the precision matrices with nonzero entries concentrated in the middle of the matrix corresponding to the evening hours. Due to the intraday serial correlations, we expect more correlations in the precision matrix, so the estimates of $\mathbf{\Omega}$ using the MRCE methods might be too simple to capture the conditional dependency structure of the electricity prices.

7.5.2 Predicting Asset Returns

We apply the MRCE method to the weekly log returns of eight US stocks from 2004 to 2010. The dataset consists of 365 observations, Figure 7.3 shows the time series

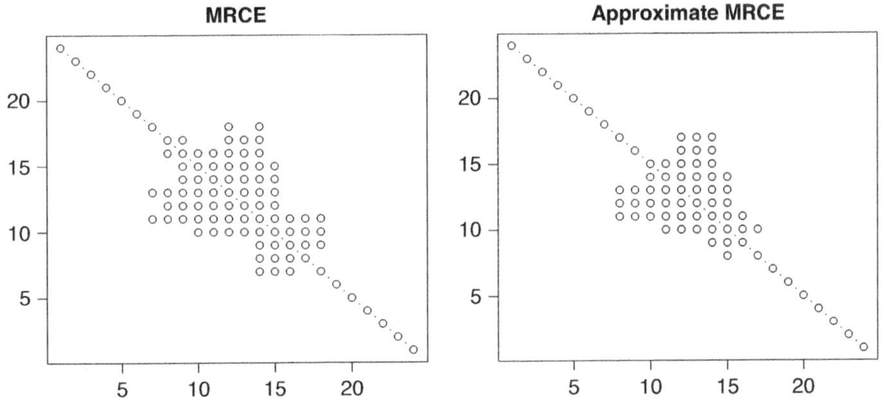

FIGURE 7.2 Positions of nonzero entries in $\widehat{\mathbf{\Omega}}$ for two methods applied to the electricity prices; the straight line indicates the diagonal of $\widehat{\mathbf{\Omega}}$.

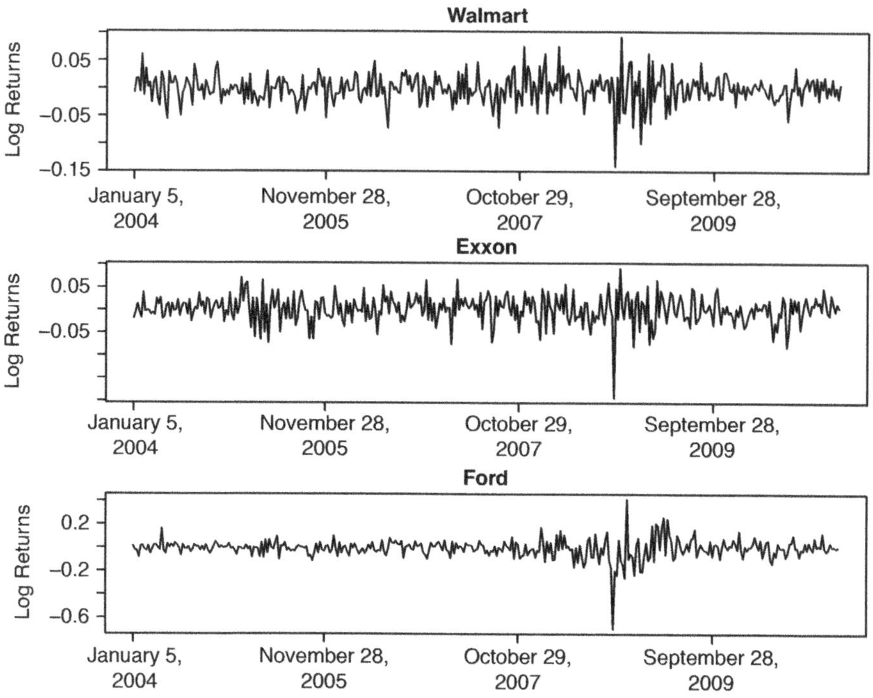

FIGURE 7.3 Plot for first three stocks.

plots of the first three stocks where reasonable similarities are seen in the dependence and volatility of the returns even though these stocks are from different sectors of the economy.

Let Y_t be the vector of log returns of the eight stocks at time t, we fit a first-order vector autoregressive (VAR) model to the data which written in matrix form becomes

$$Y = X\mathbf{B} + \mathbf{E},$$

with $Y = (Y_2', Y_3', \ldots, Y_T')'$, $X = (Y_1', Y_2', \ldots, Y_{T-1}')'$, and $\mathbf{B}$ is the transition matrix. This is a special case of the multivariate linear regression model.

To assess the predictive performance of this model, we retain the log returns of all stocks for the last 26 weeks in 2010 as the test set and the remaining 339 observations as the training set. The parameters are estimated using the training set, while the predictive performance of the model is measured by the average mean squared error over the test set for each stock. The tuning parameters are selected from the set $\Lambda = \{2^x : x = -25, -24, \ldots, 10\}$ using fivefold cross-validation. We note that different partitioning of the data in the cross-validation would lead to slightly different results, here we fix a partition and present the results for illustration.

TABLE 7.2 Sparse estimated coefficient matrix B using MRCE

	Wal	Exx	Ford	GE	CPhi	Citi	IBM	AIG
Walmart	−0.0186	0	0	0	0	0	0	0
Exxon	−0.0318	0	0	0	0	0	0	0
Ford	0	−0.0014	0	0.0055	0	0	0	0
GE	0	0	0	0	0	0	0	0
CPhi	0	0	0	0	0	0	0	0
Citigroup	−0.0023	−0.0034	−0.0010	0	0	0	−0.0306	0
IBM	0	0	0	0	0	0	0	0
AIG	−0.0004	−0.0115	0	0	0.0305	0	−0.0074	0.1079

The estimated coefficient matrix **B** using MRCE is reported in Table 7.2, the approximate MRCE also identifies the same nonzero entries. It suggests that, for example, the log returns for AIG at week $t-1$ as a relevant predictor of Walmart, Exxon, CPhi, IBM, and AIG at week t. The estimated precision matrix for MRCE reported in Table 7.3 has 30 zero entries with a similar result for the approximate MRCE method.

We present the average squared prediction error for each company over the last 26 observations in 2010 in Table 7.4 where the results using the OLS estimate are also included for comparison. The MRCE and approximate MRCE methods have very close results and perform much better than the OLS in terms of the prediction error.

7.6 FURTHER READING

The framework of multivariate regression is general enough to include virtually all linear time series models, PCA, SVD, factor analysis, linear discriminant analysis, and CCA (Reinsel and Velu, 1998; Anderson, 2003; Johnson and Wichern, 2008; Hastie et al., 2009). Successful regularization methodologies in this context can have far-reaching consequences for the analysis of high-dimensional data. RRR and the related regularization ideas in Rothman et al. (2010b), Chen et al. (2012), and Chen and Huang (2013) point out to promising directions for future developments.

TABLE 7.3 Sparse precision matrix estimate using MRCE

	Wal	Exx	Ford	GE	CPhi	Citi	IBM	AIG
Walmart	1405.9	0	0	0	0	0	0	0
Exxon	0	1037.0	−1.4	0	−34.7	0	0	0
Ford	0	−1.4	178.7	−4.7	−44.1	−47.0	−47.3	−2.7
GE	0	0	−4.7	580.5	0	−107.1	0	−15.4
ConocoPhillips	0	−34.7	−44.1	0	568.2	−22.0	0	−5.5
Citigroup	0	0	−47.0	−107.1	−22.0	163.2	−16.2	−49.7
IBM	0	0	−47.3	0	0	−16.2	999.2	0
AIG	0	0	−2.7	−15.4	−5.5	−49.7	0	92.7

TABLE 7.4 Average squared prediction error for each company × 10^3 based on 26 points. Standard errors are reported in parenthesis.

	OLS	MRCE	ap. MRCE
Walmart	0.25(0.07)	0.24(0.07)	0.24(0.07)
Exxon	0.48(0.13)	0.42(0.11)	0.42(0.11)
Ford	3.17(1.01)	2.30(0.85)	2.30(0.85)
GE	1.57(0.35)	1.50(0.33)	1.50(0.33)
ConocoPhillips	0.69(0.17)	0.65(0.15)	0.65(0.15)
Citigroup	2.95(0.72)	1.72(0.44)	1.72(0.44)
IBM	0.27(0.07)	0.24(0.06)	0.24(0.06)
AIG	4.72(1.17)	3.52(0.76)	3.52(0.77)
AVE	1.76(0.24)	1.32(0.18)	1.32(0.18)

PROBLEMS

1. (a) Using the expression in (7.2) show that

$$Q_{\boldsymbol{\Sigma}}(\boldsymbol{B}) = Q_{\boldsymbol{\Sigma}}(\widehat{\boldsymbol{B}}) + \mathrm{tr}\boldsymbol{\Sigma}^{-1}(\widehat{\boldsymbol{B}} - \boldsymbol{B})'\boldsymbol{X}'\boldsymbol{X}(\widehat{\boldsymbol{B}} - \boldsymbol{B}),$$

where $\widehat{\boldsymbol{B}} = \widehat{\boldsymbol{B}}_{\mathrm{OLS}}$.

(b) Use the identity in (a) to conclude that $Q_{\boldsymbol{\Sigma}}(\boldsymbol{B})$ is minimized at $\boldsymbol{B} = \widehat{\boldsymbol{B}}_{\mathrm{OLS}}$.

2. Let $(\boldsymbol{X}', \boldsymbol{Y}')'$ be a $(p+q)$-dimensional zero-mean random vector with a partitioned covariance matrix

$$\boldsymbol{\Sigma} = \begin{pmatrix} \boldsymbol{\Sigma}_{11} & \boldsymbol{\Sigma}_{12} \\ \boldsymbol{\Sigma}_{21} & \boldsymbol{\Sigma}_{22} \end{pmatrix},$$

where $\boldsymbol{\Sigma}_{11}$ is positive definite. For a general positive-definite matrix $\boldsymbol{W}$, its symmetric square root is denoted by $\boldsymbol{W}^{1/2}$.

(a) Let $\boldsymbol{C}$ be an arbitrary $q \times p$ matrix. Show that

$$E(\boldsymbol{Y} - \boldsymbol{C}\boldsymbol{X})'\boldsymbol{W}(\boldsymbol{Y} - \boldsymbol{C}\boldsymbol{X}) = \mathrm{tr}\{(\boldsymbol{\Sigma}^*_{22} - \boldsymbol{\Sigma}^*_{21}\boldsymbol{\Sigma}_{11}^{-1}\boldsymbol{\Sigma}^*_{12} + \boldsymbol{R}(\boldsymbol{C})\boldsymbol{R}(\boldsymbol{C})'\},$$

where $\boldsymbol{\Sigma}^*_{22} = \boldsymbol{W}^{1/2}\boldsymbol{\Sigma}_{22}\boldsymbol{W}^{1/2}$, $\boldsymbol{\Sigma}^*_{12} = \boldsymbol{\Sigma}_{12}\boldsymbol{W}^{1/2}$, $\boldsymbol{C}^* = \boldsymbol{W}^{1/2}\boldsymbol{C}$, and

$$\boldsymbol{R}(\boldsymbol{C}) = \boldsymbol{C}^*\boldsymbol{\Sigma}_{11}^{1/2} - \boldsymbol{\Sigma}^*_{21}\boldsymbol{\Sigma}_{11}^{-1/2}.$$

(b) Set $Q(\boldsymbol{C}) = E(\boldsymbol{Y} - \boldsymbol{C}\boldsymbol{X})'\boldsymbol{W}(\boldsymbol{Y} - \boldsymbol{C}\boldsymbol{X})$. Show that $Q(\cdot)$ is minimized by

$$\boldsymbol{C}_0 = \boldsymbol{\Sigma}_{12}\boldsymbol{W}^{1/2}\boldsymbol{\Sigma}_{11}^{-1},$$

which does not depend on $\boldsymbol{W}$.

Find the minimum value of $Q(\cdot)$. Interpret the task of minimizing this function from the point of view of predicting the vector $\boldsymbol{Y}$ given $\boldsymbol{X}$ and their covariance matrix $\boldsymbol{\Sigma}$.

(c) Now let $\boldsymbol{C}$ be a $q \times p$ matrix of rank r. Use the Eckart–Young Theorem and show that $Q(\cdot)$ is minimized by setting

$$\boldsymbol{C}^{*}\boldsymbol{\Sigma}^{1/2} = \sum_{j=1}^{r} \lambda^{1/2}\boldsymbol{u}_j\mathbf{v}'_j,$$

where $\boldsymbol{u}_j$ is the eigenvector associated with the jth largest eigenvalue λ_j of the matrix

$$\boldsymbol{\Sigma}_{21}^{*}\boldsymbol{\Sigma}_{11}^{-1}\boldsymbol{\Sigma}_{12}^{*} = \boldsymbol{W}^{1/2}\boldsymbol{\Sigma}_{21}\boldsymbol{\Sigma}_{11}^{-1}\boldsymbol{\Sigma}_{12}\boldsymbol{W}^{1/2},$$

and

$$\mathbf{v}_j = \lambda_j^{-1/2}\boldsymbol{\Sigma}_{11}^{-1/2}\boldsymbol{\Sigma}_{12}^{*}\boldsymbol{u}_j = \lambda_j^{-1/2}\boldsymbol{\Sigma}_{11}^{-1/2}\boldsymbol{\Sigma}_{12}\boldsymbol{W}^{1/2}\boldsymbol{u}_j.$$

Find the minimum value of the function $Q(\cdot)$.

(d) Use the result in (c) and show that the rank r matrix minimizing $Q(\cdot)$ is given by

$$\boldsymbol{C}_{(r)} = \boldsymbol{W}^{-1/2}\left(\sum_{j=1}^{r} \boldsymbol{u}_j\boldsymbol{u}'_j\right)\boldsymbol{W}^{1/2}\boldsymbol{\Sigma}_{21}\boldsymbol{\Sigma}_{11}^{-1},$$

and note that it does depend on $\boldsymbol{W}$.

3. Let $\boldsymbol{C} = \boldsymbol{C}_1\boldsymbol{C}_2$ where $\boldsymbol{C}_1$ and $\boldsymbol{C}_2$ are, respectively, $q \times r$ and $r \times p$ matrices of full-rank r.

(a) Show that $Q(\cdot)$ is minimized by the following rank r matrices:

$$\boldsymbol{C}_1(r) = \boldsymbol{W}^{-1/2}\boldsymbol{U}_{(r)},\ \boldsymbol{C}_2(r) = \boldsymbol{U}'_{(r)}\boldsymbol{W}^{1/2}\boldsymbol{\Sigma}_{21}\boldsymbol{\Sigma}_{11}^{-1},$$

where $\boldsymbol{U}_{(r)} = (\boldsymbol{u}_1, \ldots, \mathrm{U}_r)$ with the jth column $\boldsymbol{u}_j$ the eigenvector defined in the previous problem.

(b) Show that

$$\boldsymbol{C}_{(r)} = \boldsymbol{P}(\boldsymbol{W}, r)\boldsymbol{C}_{(m)},$$

where

$$P(W,r) = W^{-1/2}\left(\sum_{j=1}^{r} u_j u_j'\right) W^{1/2},$$

is an idempotent, asymmetric matrix in general.

4. **Connection with PCA:** In the above formulation, show that setting $X = Y$, $p = q$, and $W = I_p$, then Problem 1 reduces to the PCA discussed in Chapter 3.

5. **Connection with Canonical Correlation Analysis (CCA):** Set

$$W = \Sigma_{22}^{-1},$$

assuming that it is invertible. Then,

(a) Show that the above framework reduces to the CCA discussed in Chapter 7, see also Johnson and Wichern (2008).

(b) If Y is a vector of binary variables indicating class or group membership, then the RRR reduces to Fisher's *linear discriminant analysis* (Johnson and Wichern, 2008).

(c) If both X and Y are vectors of binary variables indicating the rows and columns of a two-way contingency table to which an observation belongs, the RRR reduces to *correspondence analysis* (Johnson and Wichern, 2008).

BIBLIOGRAPHY

Aldrin M. (2000). Prediction using softly shrunk reduced-rank regression. *Am Stat*, 54, 29–34.

Allen G, Tibshirani R. (2010). Transposable regularized covariance models with an application to missing data imputation. *Ann Appl Stat*, 4, 764–790.

Anderson T. (1951). Estimating linear restrictions on regression coefficients for multivariate normal distributions. *Ann Math Stat*, 22, 327–351.

— (1973). Asymptotically efficient estimation of covariance matrices with linear structure. *Ann Stat*, 1, 135–141.

— (2003). *An Introduction to Multivariate Statistics*. New York: Wiley.

Bai Z, Silverstein J. (2010). *Spectral Analysis of Large Dimensional Random Matrices*. New York: Springer.

Banerjee O, Ghaoui LE, d'Aspermont A. (2008). Model selection through sparse maximum likelihood estimation for multivariate Gaussian or binary data. *J Mach Learn Res*, 9, 485–516.

Barnard J, McCulloch R, Meng X. (2000). Modeling covariance matrices in terms of standard deviations and correlations, with applications to shrinkage. *Stat Sinica*, 10, 1281–1312.

Bartlett MS. (1933). On the theory of statistical regression. *P Roy Soc Edinb*, 53, 260–283.

Belloni A, Chernozhukov V, Wag L. (2012). Square-root lasso: Pivotal recovery of sparse signals via conic programming. *Biometrika*, 98, 791–806.

Bhatia H. (2006). Infinitely divisible matrices. *Am Math Mon*, 113, 221–235.

Bickel P, Gel YR. (2011). Banded regularization of autocovariance matrices in application to parameter estimation and forecasting of time series. *J Roy Stat Soc B*, 73, 711–728.

High-Dimensional Covariance Estimation, First Edition. Mohsen Pourahmadi.

Bickel P, Levina L. (2004). Some theory for Fisher's linear discriminant function, naive Bayes and some alternatives when there are many more variables than observations. *Bernoulli*, 10, 989–1010.

— (2008a). Regularized estimation of large covariance matrices. *Ann Stat*, 36, 199–227.

Bickel P, Levina E. (2008b). Covariance regularization by thresholding. *Ann Stat*, 36, 2577–2604.

Bickel P, Li B. (2006). Regularization in statistics. *Test*, 15, 271–344.

Bien J, Tibshirani R. (2011). Sparse estimation of a covariance matrix. *Biometrika*, 98, 807–820.

Böhm H, von Sachs R. (2008). Structural shrinkage of nonparametric spectral estimators for multivariate time series. *Electron J Stat*, 2, 696–721.

Boyd S, Vandenberghe L. (2011). *Convex Optimization*. Cambridge: Cambridge University Press.

Breiman L. (1996). Heuristics of instability and stabilization in model selection. *Ann Stat*, 24, 2350–2383.

Breiman L, Friedman J. (1997). Predicting multivariate responses in multiple linear regression. *J Roy Stat Soc Ser B*, 59, 3–37.

Brown P, Le N, Zidek J. (1994). Infer for a covariance matrix. In: Freeman PR, Smith AFM, editors. *Aspects of Uncertainty*. Chichester, UK: Wiley.

Bühlmann P, van de Geer S. (2011). *Statistics for High-Dimensional Data: Methods, Theory and Applications*. Heidelberg, Germany: Springer.

Butte AJ, Tamayo P, Slonim D, Golub TR, Kohane IS. (2000). Discovering functional relationships between RNA expression and chemotherapeutic susceptibility using relevance networks. *Proc Natl Acad Sci USA*, 97, 12182–12186.

Cadima J, Jolliffe IT. (1995). Loadings and correlations in the interpretation of principal components. *J Appl Stat*, 22, 203–214.

Cai TT. Zhang, C-H, Zhou HH. (2010). Optimal rates of convergence for covariance matrix estimation. *Ann Stat*, 38, 2118–2144.

Cai TT, Jiang T. (2011). Limiting laws of coherence of random matrices with applications to testing covariance structure and construction of compressed sensing matrices. *Ann Stat*, 39, 1496–1525.

Cai TT, Liu W. (2011). Adaptive thresholding for sparse covariance estimation. *J Am Stat Assoc*, 106, 672–684.

Cai TT, Liu W, Luo X. (2011). A constrained ℓ_1 minimization approach to sparse precision matrix estimation. *J Am Stat Assoc*, 106, 594–607.

Candes E, Tao T. (2007). The dantzig selector: statistical estimation when p is much larger than n. *Ann Stat*, 35, 2313–2351.

Carroll R, Ruppert D. (1988). *Transformation and Weighting in Regression*. London: Chapman and Hall.

Chamberlain G, Rothschild M. (1983). Arbitrage, factor structure and meanvariance analysis in large asset markets. *Econometrica*, 51, 1305–1324.

Chang C, Tsay R. (2010). Estimation of covariance matrix via the sparse cholesky factor with Lasso. *J Stat Plan Infer*, 140, 3858–3873.

Chen K, Chan KS, Chr. Stenseth NC. (2012). Reduced rank stochastic regression with a sparse singular value decomposition. *J Roy Stat Soc B*, 74, 203–221.

Chen L, Huang JZ. (2013). Sparse reduced rank regression for simultaneous dimension reduction and variable selection. *J Am Stat Assoc*, 107, 1533–1545.

Chen Z, Dunson D. (2003). Random effects selection in linear mixed models. *Biometrics*, 59, 762–769.

Chiu T, Leonard T, Tsui K. (1996). The matrix-logarithm covariance model. *J Am Stat Assoc*, 91, 198–210.

Cressie N. (2003). *Statistics for Spatial Data*. Revised ed. New York: Wiley.

Dai M, Guo W. (2004). Multivariate spectral analysis using Cholesky decomposition. *Biometrika*, 1, 629–643.

Danaher P, Wang P, Witten D. (2011). The joint graphical Lasso for inverse covariance estimation across multiple classes. Available at http://arxiv.org/abs/1111.0324.

Daniels M. (2005). A class of shrinkage priors for the dependence structure in longitudinal data. *J Stat Plan Infer*, 127, 119–130.

Daniels M, Kass R. (1999). Nonconjugate Bayesian estimation of covariance matrices in hierarchical models. *J Am Stat Assoc*, 94, 1254–1263.

Daniels M, Pourahmadi M. (2002). Dynamic models and Bayesian analysis of covariance matrices in longitudinal data. *Biometrika*, 89, 553–566.

Daniels MJ, Kass RE. (2001). Shrinkage estimators for covariance matrices. *Biometrics*, 57, 1173–1184.

d'Aspremont A, Banerjee O, El Ghaoui L. (2008). First-order methods for sparse covariance selection. *SIAM J Matrix Anal Appl*, 30, 56–66.

Dégerine S, Lambert-Lacroix S. (2003). Partial autocorrelation function of a nonstationary time series. *J Multivariate Anal*, 89, 135–147.

Dempster A. (1972). Covariance selection models. *Biometrics*, 28, 157–175.

Dempster AP, Laird NM, Rubin DB. (1977). Maximum likelihood from incomplete data via the EM algorithm. *J Roy Stat Soc B Met*, 39, 1–38.

Deng X, Tsui KW. (2013). Penalized covariance matrix estimation using a matrix-logarithm transformation. *J Comput Stat Graph*, in press.

Dey D, Srinivasan C. (1985). Estimation of a covariance matrix under Stein's loss. *Ann Stat*, 13, 1581–1591.

Domingos P, Pazzani M. (1997). On the optimality of the simple Bayesian classifier under zero-one loss. *Mach Learn*, 29, 103–130.

Donoho D. (2000). *Aide-Memoire*. High-dimensional data analysis: The curses and blessings of dimensionality. American Mathematical Society. Available at http://www.stat.stanford.edu/~donoho/Lectures/AMS2000/AMS2000.html.

— (2006). Compressed sensing. *IEEE Inform Theory*, 52, 1289–1306.

Donoho D, Johnstone I. (1994). Ideal spatial adapatation via wavelet shrinkage. *Biometrika*, 81, 425–455.

Drton M, Perlman M. (2004). Model selection for Gaussian concentration graphs. *Biometrika*, 91, 591–602.

Dudoit S, Fridlyand J, Speed T. (2002). Comparison of discrimination methods for the classification of tumors using gene expression data. *J Am Stat Assoc*, 97, 77–87.

Eaves D, Chang T. (1992). Priors for ordered conditional variances and vector partial correlation. *J Multivariate Anal*, 41, 43–55.

Efron B. (2010). *Large-Scale Inference: Empirical Bayes Methods for Estimation, Testing, and Prediction*. Cambridge: Cambridge University Press.

Efron B, Hastie T, Johnstone I, Tibshirani R. (2004). Least angle regression. *Ann Stat*, 32, 407–499.

El Karoui N. (2008). Operator norm consistent estimation of large dimensional sparse covariance matrices. *Ann Stat*, 36, 2712–2756.

Fama E, French KR. (1992). The cross-section of expected stock returns. *J Financ*, 47, 427–465.

Fan J, Feng Y, Wu Y. (2009). Network exploration via the adaptive LASSO and SCAD penalties. *Ann Appl Stat*, 3, 521–541.

Fan J, Li R. (2001). Variable selection via nonconcave penalized likelihood and its oracle properties. *J Am Stat Assoc*, 96, 1348–1360.

Fan J, Liao Y, Mincheva M. (2013). Large covariance estimation by thresholding principal orthogonal complements. *J Roy Stat Soc B*, in press.

Fan J, Lv J. (2008). Sure independence screening for ultrahigh dimensional feature space. *J Roy Stat Soc B*, 70, 849–911.

Feuerverger A, He Y, Khatri SJ. (2011). Statistical significance of the Netflix Challenge. *Stat Sci*, 27, 202–231.

Fiecas M, Ombao H. (2011). The generalized shrinkage estimator for the analysis of functional connectivity of brain signals. *Ann Appl Stat*, 5, 1102–1125.

Fox E, Dunson D. (2011). Bayesian nonparametric covariance regression. Available at http://arxiv.org/abs/1101.2017.

Friedman J. (1989). Regularized discriminant analysis. *J Am Stat Assoc*, 84, 165–175.

Friedman J, Hastie T, Tibshirani R. (2008). Sparse inverse covariance estimation with the graphical Lasso. *Biostatistics*, 9, 432–441.

— (2010). Applications of the Lasso and grouped Lasso to the estimation of sparse graphical models. Technical Report, Stanford University.

Furrer R, Bengtsson T. (2007). Estimation of high-dimensional prior and posterior covariance matrices in Kalman filter variants. *J Multivariate Anal*, 98, 227–255.

Gabriel KR. (1962). Ante-dependence analysis of an ordered set of variables. *Ann Math Stat*, 33, 201–212.

Gabriel KR, Zamir S. (1979). Lower rank approximation of matrices by least squares with any choice of weights. *Technometrics*, 21, 489–498.

Garcia T, Kohli P, Pourahmadi M. (2012). Regressograms and mean-covariance models for incomplete longitudinal data. *Am Stat*, 66, 85–91.

Geweke J, Zhou G. (1996). Measuring the pricing error of the arbitrage pricing theory. *Rev Financ Stud*, 9, 557–587.

Golub G, Van Loan C. (1996). *Matrix Computations*. Baltimore, MD: The Johns Hopkins University Press.

Guo J, Levina E, Michailidis G, Zhu J. (2011). Joint estimation of multiple graphical models. *Biometrika*, 98, 432–441.

Haff LR. (1980). Empirical Bayes estimation of the multivariate normal covariance matrix. *Ann Stat*, 8, 586–597.

— (1991). The variational form of certain Bayes estimators. *Ann Stat*, 19, 1163–1190.

Hastie T, Tibshirani R, Frieman J. (2009). *The Elements of Statistical Learning: Data Mining, Inference, and Prediction*. 2nd ed. New York: Springer.

Hoerl AE, Kennard R. (1970). Ridge regression: biased estimation for nonorthogonal problems. *Technometrics*, 12, 55–67.

Hoff P. (2009). A hierarchical eigenmodel for pooled covariance estimation. *J Roy Stat Soc B*, 71, 971–992.

Hoff P, Niu X. (2011). A covariance regression model, Technical. Report. University of Washington. Available at http://arxiv.org/abs/1102.5721.

Holmes RB. (1991). On random correlation matrices. *SIAM J Matrix Anal Appl*, 12, 239–272.

Hotelling H. (1935). The most predictable criterion. *J Educ Psychol*, 26, 139–143.

Huang J, Chen M, Maadooliat M, Pourahmadi M. (2012). A cautionary note on generalized linear models for covariance of unbalanced longitudinal date. *J Stat Plan Infer*, 142, 743–751.

Huang J, Liu L, Liu N. (2007). Estimation of large covariance matrices of longitudinal data with basis function approximations. *J Stat Comput Graphics*, 16, 189–209.

Huang J, Liu N, Pourahmadi M, Liu L. (2006). Covariance matrix selection and estimation via penalized normal likelihood. *Biometrika*, 93, 85–98.

Huang J, Shen H, Buja A. (2009). The analysis of two-way functional data using two-way regularized singular value decompositions. *J Am Stat Assoc*, 104, 1609–1620.

Izenman A. (1975). Reduced-rank regression for the multivariate linear model. *J Multivariate Anal*, 84, 707–716.

Jeffers JNR. (1967). Two case studies in the application of principal component analysis. *J Roy Stat Soc C*, 16, 225–236.

Jiang G, Sarkar SK, Hsuan F. (1999). A likelihood ratio test and its modifications for the homogeneity of the covariance matrices of dependent multivariate normals. *J Stat Plan Infer*, 81, 95–111.

Joe H. (2006). Generating random correlation matrices based on partial correlations. *J Multivariate Anal*, 97, 2177–2189.

Johnson R, Wichern D. (2008). *Applied Multivariate Statistical Analysis*. Upper Saddle River, New Jersey: Pearson Prentice Hall.

Johnstone I. (2001). On the distribution of the largest eigenvalue in principal component analysis. *Ann Stat*, 29, 295–327.

—(2011). *Gaussian Estimation: Sequence and Multiresolution Models*. Available at http//www.stat.stanford.edu/~imj/Book030811.pdf.

Johnstone IM, Lu AY. (2009). Sparse principal components analysis. *J Am Stat Assoc*, 104, 682–693.

Jolliffe IT, Trendafilov NT, Uddin M. (2003). A modified principal component technique based on the LASSO. *J Comput Graph Stat*, 12, 531–547.

Jolliffe IT, Uddin M. (2000). The simplified component technique: an alternative to rotated principal components. *J Comput Graph Stat*, 9, 689–710.

Jones M. (1987). Randomly choosing parameters from the stationarity and invertibility region of autoregressive-moving average models. *Appl Stat*, 36, 134–138.

Jones R. (1980). Maximum likelihood fitting of ARMA models to time series with missing observations. *Technometrics*, 22, 389–395.

Jong J, Kotz S. (1999). On a relation between principal components and regression analysis. *Am Stat*, 53, 349–351.

Kalman RE. (1960). A new approach to the linear filtering and prediction problems. *Trans ASME J Basic Eng*, 82, 35–45.

Kurowicka D, Cooke R. (2003). A parameterization of positive definite matrices in terms of partial correlation vines. *Linear Algebra Appl*, 372, 225–251.

Lam C, Fan J. (2009). Sparsistency and rates of convergence in large covariance matrices estimation. *Ann Stat*, 37, 4254–4278.

Ledoit O, Wolf M. (2004). A well-conditioned estimator for large-dimensional covariance matrices. *J Multivariate Anal*, 88, 365–411.

Lee M, Shen H, Huang JZ, Marron JS. (2010). Biclustering via sparse singular value decomposition. *Biometrics*, 66, 1087–1095.

Leng C, Li B. (2011). Forward adaptive banding for estimating large covariance matrices. *Biometrika*, 98, 821–830.

Leng C, Zhang W, Pan J. (2010). Semiparametric mean-covariance regression analysis for longitudinal data. *J Am Stat Assoc*, 105, 181–193.

Leonard T, Hsu J. (1992). Bayesian inference for a covariance matrix. *Ann Stat*, 20, 1669–1696.

LeSage JP, Pace RK. (2007). A matrix exponential spatial specification. *J Econometrics*, 140, 198–214.

Levina E, Rothman A, Zhu J. (2008). Sparse estimation of large covariance matrices via a nested Lasso penalty. *Ann Appl Stat*, 2, 245–263.

Li L, Toh KC. (2010). An inexact interior point method for L_1-regularized sparse covariance selection. *Math Prog Comp*, 2, 291–315.

Liang K, Zeger S. (1986). Longitudinal data analysis using generalized linear models. *Biometrika*, 73, 13–22.

Lin S, Perlman M. (1985). A Monte Carlo comparison of four estimators of a covariance matrix. In: Krishnaiah PR, editor. *Multivariate Analysis*. 6th ed. Amsterdam: North-Holland.

Lindquist M. (2008). The statistical analysis of fMRI data. *Stat Sci*, 23, 439–464.

Liu C. (1993). Bartlett's decomposition of the posterior distribution of the covariance for normal monotone ignorable missing data. *J Multivariate Anal*, 46, 198–206.

Liu H, Wang L. (2012). TIGER: A tuning-insensitive approach for optimally estimating Gaussian graphical models. (arXiv:1209.2437v1).

Ma Z. (2011). Sparse principal component analysis and iterative thresholding. Available at http://arxiv.org/abs/1112.2432.

Marčenko V, Pastur L. (1967). Distribution for some sets of random matrices. *Math USSR-Sb*, 1, 457–483.

McMurray T, Politis D. (2010). Banded and tapered estimates for autocovariance matrices and the linear bootstrap. *J Time Series Anal*, 31, 471–482.

Meinshausen N, Bühlmann P. (2006). High-dimensional graphs and variable selection with the Lasso. *Ann Stat*, 34, 1436–1462.

Nadler B. (2009). Discussion of "Sparse principal components analysis", by Johnstone and Lu. *J Am Stat Assoc*, 104, 694–697.

Oakes D. (1999). Direct calculation of the information matrix via the EM algorithm. *J Roy Stat Soc B*, 61, 479–482.

Pan J, MacKenzie G. (2003). On modelling mean-covariance structure in longitudinal studies. *Biometrika*, 90, 239–244.

Panagiotelis A, Smith M. (2008). Bayesian density forecasting of intraday electricity prices using multivariate skew t distributions. *Int J Forecasting*, 24, 710–727.

Paul D. (2007). Asymptotics of sample eigenstructure for large dimensional spiked covariance model. *Stat Sinica*, 17, 1617–1642.

Peng J, Wang P, Zhou N, Zhu J. (2009). Partial correlation estimation by joint sparse regression models. *J Am Stat Assoc*, 104, 735–746.

Peng J, Zhu J, Bergamaschi A, Han W, Noh DY, Pollack J, Wang P. (2010). Regularized multivariate regression for identifying master predictors with applicaions to integrative genomics study of breast cancer. *Ann Appl. Stat*, 4, 53–77.

Pourahmadi M. (1999). Joint mean-covariance models with applications to longitudinal data: Unconstrained parameterisation. *Biometrika*, 86, 677–690.

— (2000). Maximum likelihood estimation of generalized linear models for multivariate normal covariance matrix. *Biometrika*, 87, 425–435.

— (2001). *Foundations of Time Series Anal and Prediction Theory*. New York: Wiley.

— (2004). Simultaneous modeling of covariance matrices: GLM, Bayesian and nonparametric perspective. *Correlated Data Modelling 2004, D. Gregori et al. (eds). FrancoAngeli, Milan, Italy.*

— (2007a). Cholesky decompositions and estimation of a multivariate normal covariance matrix: Parameter orthogonality. *Biometrika*, 94, 1006–1013.

— (2007b). Simultaneous modelling of covariance matrices: GLM, Bayesian and nonparametric perspective. *Correlated Data Modelling 2004, D. Gregori et al. (eds.). FrancoAngeli, Milan, Italy.*

— (2011). Covariance estimation: The GLM and regularization perspectives. *Stat Sci*, 26, 369–387.

Quenouille MH. (1949). Approximate tests of correlation in time series. *J Roy Stat Soc B*, 11, 68–84.

Rajaratnam B, Massam H, Corvallo C. (2008). Flexible covariance estimation in graphical Gaussian models. *Ann Stat*, 36, 2818–2849.

Ramsey FL. (1974). Characterization of the partial autocorrelation function. *Ann Stat*, 2, 1296–1301.

Ramsey J, Silverman B. (2005). *Functional Data Analysis* 2nd ed, New York: Springer.

Ravikumar P, Wainwright MJ, Yu B. (2011). High-dimensional covariance estimation by minimizing ℓ_1-penalized log-determinant divergence. *Electron J Stat*, 5, 935–980.

Reinsel G, Velu R. (1998). *Multivariate Reduced-Rank Regression: Theory Applications*. New York: Springer.

Rocha GV, Zhao P, Yu B. (2008). A path following algorithm for sparse pseudolikelihood inverse covariance estimation (splice). Technical Report No. 759, Statistics Department, University of California at Berkeley.

Rothman A. (2012). Positive definite estimators of large covariance matrices. *Biometrika*, 99, 733–740.

Rothman A, Bickel P, Levina E, Zhu J. (2008). Sparse permutation invariant covariance estimation. *Electron J Staistics*, 2, 494–515.

Rothman A, Levina E, Zhu J. (2010a). A new approach to Cholesky-based estimation of high-dimensional covariance matrices. *Biometrika*, 97, 539–550.

— (2010b). Sparse multivariate regression with covariance estimation. *J Comput Graph Stat*, 19, 947–962.

Rothman AJ, Levina E, Zhu J. (2009). Generalized thresholding of large covariance matrices. *J Am Stat Assoc*, 104, 177–186.

Roy J. (1958). Step-down procedure in multivariate-analysis. *Ann Math Statist*, 29, 1177–1187

Schott JR. (2012). A note on maximum likelihood estimation for covariance reducing models. *Stat and Probabil Lett*, 82, 1629–1631.

Searle S, Casella G, McCulloch C. (1992). *Variance Components*. New York: Wiley.

Sharpe W. (1970). *Protfolio Theory and Capital Markets*. New York: McGraw-Hill.

Shen H, Huang J. (2008). Sparse principal component analysis via regularized low rank matrix approximation. *J Multivariate Anal*, 99, 1015–1034.

Simon B. (2005). *Othogonal Polynomials on the Unit Circle, Part I and II*. Providence, RI: American Mathematical Society.

Smith M, Kohn R. (2002). Parsimonious covariance matrix estimation for longitudinal data. *J Am Stat Assoc*, 97, 1141–1153.

Stein C. (1956). Inadmissibility of the usual estimator of the mean of a multivariate normal distribution. In: Neyman J, editor. *Proceedings of the Third Berkeley Sympossium on Mathematical and Statistical Probability*. University of California Press, Vol. 1, pp. 197–206.

Sun D, Sun, X. (2006). Estimation of multivariate normal precision and covariance matrices in a star-shaped model with missing date. *J Multivariate Anal*, 97, 698–719.

Szatrowski TH. (1980). Necessary and sufficient conditions for explicit solutions in the multivariate normal estimation problem for patterned means and covariances. *Ann Stat*, 8, 802–810.

Tibshirani R. (1996). Regression shrinkage and selection via the Lasso. *J Roy Stat Soc B*, 58, 267–288.

Tibshirani R. (2011). Regression shrinkage and selection via Lasso: a retrospective. *J Roy Stat*, 73, 273–282.

Vines S. (2000). Simple principal components. *Appl Stat*, 49, 441–451.

Warton D. (2008). Penalized normal likelihood and ridge regularization of correlation and covariance matrices. *J Am Stat Assoc*, 103, 340–349.

Wermuth N. (1980). Linear recursive equations, covariance selection and path analysis. *J Am Stat Assoc*, 75, 963–972.

Whittaker J. (1990). *Graphical Models in Appl Multivariate Statistics*. Chichester, UK: John Wiley and Sons.

Witten DM, Tibshirani R. (2009). Covariance-regularized regression and classification for high-dimensional problems. *J Roy Stat Soc B*, 71, 615–636.

Witten DM, Tibshirani R, Hastie T. (2009). A penalized matrix decomposition, with applications to sparse principal components and canonical correlation analysis. *Biostatistics*, 10, 515–534.

Wold HOA. (1960). A generalization of causal chain models. *Econometrica*, 28, 443–463.

Won JH, Lim J, Kim SJ, Rajaratnam B. (2009). Maximum likelihood covariance estimation with the condition number constraint. Technical Report, Department of Statistics, Stanford University.

Wong F, Carter C, Kohn R. (2003). Efficient estimation of covariance selection models. *Biometrika*, 90, 809–830.

Wright S. (1934).The method of path coefficients. *Ann Math Statist*, 5, 161–215.

Wu WB, Pourahmadi M. (2003). Nonparametric estimation of large covariance matrices of longitudinal data. *Biometrika*, 90, 831–844.

— (2009). Banding sample covariance matrices of stationary processes. *Stat Sinica*, 19, 1755–1768.

Xue L, Ma S, Zou H. (2012). Positive definite ℓ_1 penalized estimation of large covariance matrices. *J Am Stat Assoc*, 107, 1480–1491.

Yang D, Ma Z, Buja A. (2011). A sparse SVD method for high-dimensional data. Available at http://arxiv.org/abs/1112.2433.

Yang R, Berger J. (1994). Estimation of a covariance matrix using the reference prior. *Ann Statist*, 22, 1195–1211.

Ye H, Pan J. (2006). Modeling covariance structures in generalized estimating equations for longitudinal data. *Biometrika*, 93, 911–926.

Yuan M. (2010). High-dimensional inverse covariance estimation via linear programming. *J Mach Learn Res*, 11, 2261–2286.

Yuan M, Ekici A, Lu Z, Monteiro R. (2007). Dimension reduction and coefficient estimation in multivariate linear regression. *J Roy Stat Soc B*, 69, 329–346.

Yuan M, Lin Y. (2007). Model selection and estimation in the Gaussian graphical model. *Biometrika*, 94, 19–35.

Yule G. (1927). On a model of investigating periodicities in disturbed series with special reference to Wolfer's sunspot numbers. *Phil Trans A*, 226, 267–298.

Yule GU. (1907). On the theory of correlation for any number of variables, treated by a new system of notation. *P R Soc London Roy Soc*, 79, 85–96.

Zhao T, Liu H, Roeder K, Lafferty J, Wasserman L. (2012). The huge package for high-dimensional undirected graph estimation in R. *J Mach Learn Res*, 98888, 1059–1062.

Zhu Z, Liu Y. (2009). Estimating spatial covariance using penalized likelihood with weighted L1 penalty. *J Nonparametr Stat*, 21, 925–942.

Zimmerman D, Núñez Antón V. (2009). *Antedependence Models for Longitudinal Data*. Boca Raton, Florida: CRC Press.

Zimmerman DL. (2000). Viewing the correlation structure of longitudinal data through a PRISM. *Am Stat*, 54, 310–318.

Zou H. (2006). The adaptive Lasso and its oracle properties. *J Am Stat Assoc*, 101, 1418–1429.

Zou H, Hastie T. (2005). Regularization and variable selection via the Elastic Net. *J Roy Stat Soc B*, 67, 301–320.

Zou H, Hastie T, Tibshirani R. (2006). Sparse principal component analysis. *J Comput Graph Stat*, 15, 265–286.

— (2007). On the degrees of freedom of the Lasso. *Ann Stat*, 35, 2173–2192.

INDEX

Page numbers followed by f and t indicate figures and table respectively.

WILEY SERIES IN PROBABILITY AND STATISTICS

ESTABLISHED BY WALTER A. SHEWHART AND SAMUEL S. WILKS

Editors: *David J. Balding, Noel A. C. Cressie, Garrett M. Fitzmaurice, Harvey Goldstein, Iain M. Johnstone, Geert Molenberghs, David W. Scott, Adrian F. M. Smith, Ruey S. Tsay, Sanford Weisberg*
Editors Emeriti: *Vic Barnett, J. Stuart Hunter, Joseph B. Kadane, Jozef L. Teugels*

The ***Wiley Series in Probability and Statistics*** is well established and authoritative. It covers many topics of current research interest in both pure and applied statistics and probability theory. Written by leading statisticians and institutions, the titles span both state-of-the-art developments in the field and classical methods.

Reflecting the wide range of current research in statistics, the series encompasses applied, methodological and theoretical statistics, ranging from applications and new techniques made possible by advances in computerized practice to rigorous treatment of theoretical approaches.

This series provides essential and invaluable reading for all statisticians, whether in academia, industry, government, or research.

† ABRAHAM and LEDOLTER · Statistical Methods for Forecasting
AGRESTI · Analysis of Ordinal Categorical Data, *Second Edition*
AGRESTI · An Introduction to Categorical Data Analysis, *Second Edition*
AGRESTI · Categorical Data Analysis, *Second Edition*
ALTMAN, GILL, and McDONALD · Numerical Issues in Statistical Computing for the Social Scientist
AMARATUNGA and CABRERA · Exploration and Analysis of DNA Microarray and Protein Array Data
ANDĚL · Mathematics of Chance
ANDERSON · An Introduction to Multivariate Statistical Analysis, *Third Edition*
* ANDERSON · The Statistical Analysis of Time Series
ANDERSON, AUQUIER, HAUCK, OAKES, VANDAELE, and WEISBERG · Statistical Methods for Comparative Studies
ANDERSON and LOYNES · The Teaching of Practical Statistics
ARMITAGE and DAVID (editors) · Advances in Biometry
ARNOLD, BALAKRISHNAN, and NAGARAJA · Records
* ARTHANARI and DODGE · Mathematical Programming in Statistics
* BAILEY · The Elements of Stochastic Processes with Applications to the Natural Sciences
BAJORSKI · Statistics for Imaging, Optics, and Photonics
BALAKRISHNAN and KOUTRAS · Runs and Scans with Applications
BALAKRISHNAN and NG · Precedence-Type Tests and Applications
BARNETT · Comparative Statistical Inference, *Third Edition*
BARNETT · Environmental Statistics
BARNETT and LEWIS · Outliers in Statistical Data, *Third Edition*
BARTHOLOMEW, KNOTT, and MOUSTAKI · Latent Variable Models and Factor Analysis: A Unified Approach, *Third Edition*
BARTOSZYNSKI and NIEWIADOMSKA-BUGAJ · Probability and Statistical Inference, *Second Edition*
BASILEVSKY · Statistical Factor Analysis and Related Methods: Theory and Applications
BATES and WATTS · Nonlinear Regression Analysis and Its Applications
BECHHOFER, SANTNER, and GOLDSMAN · Design and Analysis of Experiments for Statistical Selection, Screening, and Multiple Comparisons

*Now available in a lower priced paperback edition in the Wiley Classics Library.
†Now available in a lower priced paperback edition in the Wiley–Interscience Paperback Series.

BEIRLANT, GOEGEBEUR, SEGERS, TEUGELS, and DE WAAL · Statistics of Extremes: Theory and Applications
BELSLEY · Conditioning Diagnostics: Collinearity and Weak Data in Regression
† BELSLEY, KUH, and WELSCH · Regression Diagnostics: Identifying Influential Data and Sources of Collinearity
BENDAT and PIERSOL · Random Data: Analysis and Measurement Procedures, *Fourth Edition*
BERNARDO and SMITH · Bayesian Theory
BHAT and MILLER · Elements of Applied Stochastic Processes, *Third Edition*
BHATTACHARYA and WAYMIRE · Stochastic Processes with Applications
BIEMER, GROVES, LYBERG, MATHIOWETZ, and SUDMAN · Measurement Errors in Surveys
BILLINGSLEY · Convergence of Probability Measures, *Second Edition*
BILLINGSLEY · Probability and Measure, *Anniversary Edition*
BIRKES and DODGE · Alternative Methods of Regression
BISGAARD and KULAHCI · Time Series Analysis and Forecasting by Example
BISWAS, DATTA, FINE, and SEGAL · Statistical Advances in the Biomedical Sciences: Clinical Trials, Epidemiology, Survival Analysis, and Bioinformatics
BLISCHKE and MURTHY (editors) · Case Studies in Reliability and Maintenance
BLISCHKE and MURTHY · Reliability: Modeling, Prediction, and Optimization
BLOOMFIELD · Fourier Analysis of Time Series: An Introduction, *Second Edition*
BOLLEN · Structural Equations with Latent Variables
BOLLEN and CURRAN · Latent Curve Models: A Structural Equation Perspective
BOROVKOV · Ergodicity and Stability of Stochastic Processes
BOSQ and BLANKE · Inference and Prediction in Large Dimensions
BOULEAU · Numerical Methods for Stochastic Processes
* BOX and TIAO · Bayesian Inference in Statistical Analysis
BOX · Improving Almost Anything, *Revised Edition*
* BOX and DRAPER · Evolutionary Operation: A Statistical Method for Process Improvement
BOX and DRAPER · Response Surfaces, Mixtures, and Ridge Analyses, *Second Edition*
BOX, HUNTER, and HUNTER · Statistics for Experimenters: Design, Innovation, and Discovery, *Second Editon*
BOX, JENKINS, and REINSEL · Time Series Analysis: Forcasting and Control, *Fourth Edition*
BOX, LUCEÑO, and PANIAGUA-QUIÑONES · Statistical Control by Monitoring and Adjustment, *Second Edition*
* BROWN and HOLLANDER · Statistics: A Biomedical Introduction
CAIROLI and DALANG · Sequential Stochastic Optimization
CASTILLO, HADI, BALAKRISHNAN, and SARABIA · Extreme Value and Related Models with Applications in Engineering and Science
CHAN · Time Series: Applications to Finance with R and S-Plus®, *Second Edition*
CHARALAMBIDES · Combinatorial Methods in Discrete Distributions
CHATTERJEE and HADI · Regression Analysis by Example, *Fourth Edition*
CHATTERJEE and HADI · Sensitivity Analysis in Linear Regression
CHERNICK · Bootstrap Methods: A Guide for Practitioners and Researchers, *Second Edition*
CHERNICK and FRIIS · Introductory Biostatistics for the Health Sciences
CHILÈS and DELFINER · Geostatistics: Modeling Spatial Uncertainty, *Second Edition*
CHOW and LIU · Design and Analysis of Clinical Trials: Concepts and Methodologies, *Second Edition*
CLARKE · Linear Models: The Theory and Application of Analysis of Variance
CLARKE and DISNEY · Probability and Random Processes: A First Course with Applications, *Second Edition*

*Now available in a lower priced paperback edition in the Wiley Classics Library.
†Now available in a lower priced paperback edition in the Wiley–Interscience Paperback Series.

* COCHRAN and COX · Experimental Designs, *Second Edition*
COLLINS and LANZA · Latent Class and Latent Transition Analysis: With Applications in the Social, Behavioral, and Health Sciences
CONGDON · Applied Bayesian Modelling
CONGDON · Bayesian Models for Categorical Data
CONGDON · Bayesian Statistical Modelling, *Second Edition*
CONOVER · Practical Nonparametric Statistics, *Third Edition*
COOK · Regression Graphics
COOK and WEISBERG · An Introduction to Regression Graphics
COOK and WEISBERG · Applied Regression Including Computing and Graphics
CORNELL · A Primer on Experiments with Mixtures
CORNELL · Experiments with Mixtures, Designs, Models, and the Analysis of Mixture Data, *Third Edition*
COX · A Handbook of Introductory Statistical Methods
CRESSIE · Statistics for Spatial Data, *Revised Edition*
CRESSIE and WIKLE · Statistics for Spatio-Temporal Data
CSÖRGŐ and HORVÁTH · Limit Theorems in Change Point Analysis
DAGPUNAR · Simulation and Monte Carlo: With Applications in Finance and MCMC
DANIEL · Applications of Statistics to Industrial Experimentation
DANIEL · Biostatistics: A Foundation for Analysis in the Health Sciences, *Eighth Edition*
* DANIEL · Fitting Equations to Data: Computer Analysis of Multifactor Data, *Second Edition*
DASU and JOHNSON · Exploratory Data Mining and Data Cleaning
DAVID and NAGARAJA · Order Statistics, *Third Edition*
* DEGROOT, FIENBERG, and KADANE · Statistics and the Law
DEL CASTILLO · Statistical Process Adjustment for Quality Control
DeMARIS · Regression with Social Data: Modeling Continuous and Limited Response Variables
DEMIDENKO · Mixed Models: Theory and Applications with R, *Second Edition*
DENISON, HOLMES, MALLICK and SMITH · Bayesian Methods for Nonlinear Classification and Regression
DETTE and STUDDEN · The Theory of Canonical Moments with Applications in Statistics, Probability, and Analysis
DEY and MUKERJEE · Fractional Factorial Plans
DILLON and GOLDSTEIN · Multivariate Analysis: Methods and Applications
* DODGE and ROMIG · Sampling Inspection Tables, *Second Edition*
* DOOB · Stochastic Processes
DOWDY, WEARDEN, and CHILKO · Statistics for Research, *Third Edition*
DRAPER and SMITH · Applied Regression Analysis, *Third Edition*
DRYDEN and MARDIA · Statistical Shape Analysis
DUDEWICZ and MISHRA · Modern Mathematical Statistics
DUNN and CLARK · Basic Statistics: A Primer for the Biomedical Sciences, *Fourth Edition*
DUPUIS and ELLIS · A Weak Convergence Approach to the Theory of Large Deviations
EDLER and KITSOS · Recent Advances in Quantitative Methods in Cancer and Human Health Risk Assessment
* ELANDT-JOHNSON and JOHNSON · Survival Models and Data Analysis
ENDERS · Applied Econometric Time Series, *Third Edition*
† ETHIER and KURTZ · Markov Processes: Characterization and Convergence
EVANS, HASTINGS, and PEACOCK · Statistical Distributions, *Third Edition*
EVERITT, LANDAU, LEESE, and STAHL · Cluster Analysis, *Fifth Edition*
FEDERER and KING · Variations on Split Plot and Split Block Experiment Designs

*Now available in a lower priced paperback edition in the Wiley Classics Library.
†Now available in a lower priced paperback edition in the Wiley–Interscience Paperback Series.

FELLER · An Introduction to Probability Theory and Its Applications, Volume I, *Third Edition*, Revised; Volume II, *Second Edition*

FITZMAURICE, LAIRD, and WARE · Applied Longitudinal Analysis, *Second Edition*

* FLEISS · The Design and Analysis of Clinical Experiments

FLEISS · Statistical Methods for Rates and Proportions, *Third Edition*

† FLEMING and HARRINGTON · Counting Processes and Survival Analysis

FUJIKOSHI, ULYANOV, and SHIMIZU · Multivariate Statistics: High-Dimensional and Large-Sample Approximations

FULLER · Introduction to Statistical Time Series, *Second Edition*

† FULLER · Measurement Error Models

GALLANT · Nonlinear Statistical Models

GEISSER · Modes of Parametric Statistical Inference

GELMAN and MENG · Applied Bayesian Modeling and Causal Inference from ncomplete-Data Perspectives

GEWEKE · Contemporary Bayesian Econometrics and Statistics

GHOSH, MUKHOPADHYAY, and SEN · Sequential Estimation

GIESBRECHT and GUMPERTZ · Planning, Construction, and Statistical Analysis of Comparative Experiments

GIFI · Nonlinear Multivariate Analysis

GIVENS and HOETING · Computational Statistics

GLASSERMAN and YAO · Monotone Structure in Discrete-Event Systems

GNANADESIKAN · Methods for Statistical Data Analysis of Multivariate Observations, *Second Edition*

GOLDSTEIN · Multilevel Statistical Models, *Fourth Edition*

GOLDSTEIN and LEWIS · Assessment: Problems, Development, and Statistical Issues

GOLDSTEIN and WOOFF · Bayes Linear Statistics

GREENWOOD and NIKULIN · A Guide to Chi-Squared Testing

GROSS, SHORTLE, THOMPSON, and HARRIS · Fundamentals of Queueing Theory, *Fourth Edition*

GROSS, SHORTLE, THOMPSON, and HARRIS · Solutions Manual to Accompany Fundamentals of Queueing Theory, *Fourth Edition*

* HAHN and SHAPIRO · Statistical Models in Engineering

HAHN and MEEKER · Statistical Intervals: A Guide for Practitioners

HALD · A History of Probability and Statistics and their Applications Before 1750

† HAMPEL · Robust Statistics: The Approach Based on Influence Functions

HARTUNG, KNAPP, and SINHA · Statistical Meta-Analysis with Applications

HEIBERGER · Computation for the Analysis of Designed Experiments

HEDAYAT and SINHA · Design and Inference in Finite Population Sampling

HEDEKER and GIBBONS · Longitudinal Data Analysis

HELLER · MACSYMA for Statisticians

HERITIER, CANTONI, COPT, and VICTORIA-FESER · Robust Methods in Biostatistics

HINKELMANN and KEMPTHORNE · Design and Analysis of Experiments, Volume 1: Introduction to Experimental Design, *Second Edition*

HINKELMANN and KEMPTHORNE · Design and Analysis of Experiments, Volume 2: Advanced Experimental Design

HINKELMANN (editor) · Design and Analysis of Experiments, Volume 3: Special Designs and Applications

HOAGLIN, MOSTELLER, and TUKEY · Fundamentals of Exploratory Analysis of Variance

* HOAGLIN, MOSTELLER, and TUKEY · Exploring Data Tables, Trends and Shapes

* HOAGLIN, MOSTELLER, and TUKEY · Understanding Robust and Exploratory Data Analysis

*Now available in a lower priced paperback edition in the Wiley Classics Library.
†Now available in a lower priced paperback edition in the Wiley–Interscience Paperback Series.

HOCHBERG and TAMHANE · Multiple Comparison Procedures
HOCKING · Methods and Applications of Linear Models: Regression and the Analysis of Variance, *Third Edition*
HOEL · Introduction to Mathematical Statistics, *Fifth Edition*
HOGG and KLUGMAN · Loss Distributions
HOLLANDER, WOLFE, and CHICKEN · Nonparametric Statistical Methods, *Third Edition*
HOSMER and LEMESHOW · Applied Logistic Regression, *Second Edition*
HOSMER, LEMESHOW, and MAY · Applied Survival Analysis: Regression Modeling of Time-to-Event Data, *Second Edition*
HUBER · Data Analysis: What Can Be Learned From the Past 50 Years
HUBER · Robust Statistics
† HUBER and RONCHETTI · Robust Statistics, *Second Edition*
HUBERTY · Applied Discriminant Analysis, *Second Edition*
HUBERTY and OLEJNIK · Applied MANOVA and Discriminant Analysis, *Second Edition*
HUITEMA · The Analysis of Covariance and Alternatives: Statistical Methods for Experiments, Quasi-Experiments, and Single-Case Studies, *Second Edition*
HUNT and KENNEDY · Financial Derivatives in Theory and Practice, *Revised Edition*
HURD and MIAMEE · Periodically Correlated Random Sequences: Spectral Theory and Practice
HUSKOVA, BERAN, and DUPAC · Collected Works of Jaroslav Hajek— with Commentary
HUZURBAZAR · Flowgraph Models for Multistate Time-to-Event Data
JACKMAN · Bayesian Analysis for the Social Sciences
† JACKSON · A User's Guide to Principle Components
JOHN · Statistical Methods in Engineering and Quality Assurance
JOHNSON · Multivariate Statistical Simulation
JOHNSON and BALAKRISHNAN · Advances in the Theory and Practice of Statistics: A Volume in Honor of Samuel Kotz
JOHNSON, KEMP, and KOTZ · Univariate Discrete Distributions, *Third Edition*
JOHNSON and KOTZ (editors) · Leading Personalities in Statistical Sciences: From the Seventeenth Century to the Present
JOHNSON, KOTZ, and BALAKRISHNAN · Continuous Univariate Distributions, Volume 1, *Second Edition*
JOHNSON, KOTZ, and BALAKRISHNAN · Continuous Univariate Distributions, Volume 2, *Second Edition*
JOHNSON, KOTZ, and BALAKRISHNAN · Discrete Multivariate Distributions
JUDGE, GRIFFITHS, HILL, LÜTKEPOHL, and LEE · The Theory and Practice of Econometrics, *Second Edition*
JUREK and MASON · Operator-Limit Distributions in Probability Theory
KADANE · Bayesian Methods and Ethics in a Clinical Trial Design
KADANE AND SCHUM · A Probabilistic Analysis of the Sacco and Vanzetti Evidence
KALBFLEISCH and PRENTICE · The Statistical Analysis of Failure Time Data, *Second Edition*
KARIYA and KURATA · Generalized Least Squares
KASS and VOS · Geometrical Foundations of Asymptotic Inference
† KAUFMAN and ROUSSEEUW · Finding Groups in Data: An Introduction to Cluster Analysis
KEDEM and FOKIANOS · Regression Models for Time Series Analysis
KENDALL, BARDEN, CARNE, and LE · Shape and Shape Theory
KHURI · Advanced Calculus with Applications in Statistics, *Second Edition*
KHURI, MATHEW, and SINHA · Statistical Tests for Mixed Linear Models
* KISH · Statistical Design for Research

*Now available in a lower priced paperback edition in the Wiley Classics Library.
†Now available in a lower priced paperback edition in the Wiley–Interscience Paperback Series.

KLEIBER and KOTZ · Statistical Size Distributions in Economics and Actuarial Sciences
KLEMELÄ · Smoothing of Multivariate Data: Density Estimation and Visualization
KLUGMAN, PANJER, and WILLMOT · Loss Models: From Data to Decisions, *Third Edition*
KLUGMAN, PANJER, and WILLMOT · Loss Models: Further Topics
KLUGMAN, PANJER, and WILLMOT · Solutions Manual to Accompany Loss Models: From Data to Decisions, *Third Edition*
KOSKI and NOBLE · Bayesian Networks: An Introduction
KOTZ, BALAKRISHNAN, and JOHNSON · Continuous Multivariate Distributions, Volume 1, *Second Edition*
KOTZ and JOHNSON (editors) · Encyclopedia of Statistical Sciences: Volumes 1 to 9 with Index
KOTZ and JOHNSON (editors) · Encyclopedia of Statistical Sciences: Supplement Volume
KOTZ, READ, and BANKS (editors) · Encyclopedia of Statistical Sciences: Update Volume 1
KOTZ, READ, and BANKS (editors) · Encyclopedia of Statistical Sciences: Update Volume 2
KOWALSKI and TU · Modern Applied U-Statistics
KRISHNAMOORTHY and MATHEW · Statistical Tolerance Regions: Theory, Applications, and Computation
KROESE, TAIMRE, and BOTEV · Handbook of Monte Carlo Methods
KROONENBERG · Applied Multiway Data Analysis
KULINSKAYA, MORGENTHALER, and STAUDTE · Meta Analysis: A Guide to Calibrating and Combining Statistical Evidence
KULKARNI and HARMAN · An Elementary Introduction to Statistical Learning Theory
KUROWICKA and COOKE · Uncertainty Analysis with High Dimensional Dependence Modelling
KVAM and VIDAKOVIC · Nonparametric Statistics with Applications to Science and Engineering
LACHIN · Biostatistical Methods: The Assessment of Relative Risks, *Second Edition*
LAD · Operational Subjective Statistical Methods: A Mathematical, Philosophical, and Historical Introduction
LAMPERTI · Probability: A Survey of the Mathematical Theory, *Second Edition*
LAWLESS · Statistical Models and Methods for Lifetime Data, *Second Edition*
LAWSON · Statistical Methods in Spatial Epidemiology, *Second Edition*
LE · Applied Categorical Data Analysis, *Second Edition*
LE · Applied Survival Analysis
LEE · Structural Equation Modeling: A Bayesian Approach
LEE and WANG · Statistical Methods for Survival Data Analysis, *Fourth Edition*
LEPAGE and BILLARD · Exploring the Limits of Bootstrap
LESSLER and KALSBEEK · Nonsampling Errors in Surveys
LEYLAND and GOLDSTEIN (editors) · Multilevel Modelling of Health Statistics
LIAO · Statistical Group Comparison
LIN · Introductory Stochastic Analysis for Finance and Insurance
LITTLE and RUBIN · Statistical Analysis with Missing Data, *Second Edition*
LLOYD · The Statistical Analysis of Categorical Data
LOWEN and TEICH · Fractal-Based Point Processes
MAGNUS and NEUDECKER · Matrix Differential Calculus with Applications in Statistics and Econometrics, *Revised Edition*
MALLER and ZHOU · Survival Analysis with Long Term Survivors
MARCHETTE · Random Graphs for Statistical Pattern Recognition
MARDIA and JUPP · Directional Statistics

*Now available in a lower priced paperback edition in the Wiley Classics Library.
†Now available in a lower priced paperback edition in the Wiley–Interscience Paperback Series.

MARKOVICH · Nonparametric Analysis of Univariate Heavy-Tailed Data: Research and Practice
MARONNA, MARTIN and YOHAI · Robust Statistics: Theory and Methods
MASON, GUNST, and HESS · Statistical Design and Analysis of Experiments with Applications to Engineering and Science, *Second Edition*
McCULLOCH, SEARLE, and NEUHAUS · Generalized, Linear, and Mixed Models, *Second Edition*
McFADDEN · Management of Data in Clinical Trials, *Second Edition*
* McLACHLAN · Discriminant Analysis and Statistical Pattern Recognition
McLACHLAN, DO, and AMBROISE · Analyzing Microarray Gene Expression Data
McLACHLAN and KRISHNAN · The EM Algorithm and Extensions, *Second Edition*
McLACHLAN and PEEL · Finite Mixture Models
McNEIL · Epidemiological Research Methods
MEEKER and ESCOBAR · Statistical Methods for Reliability Data
MEERSCHAERT and SCHEFFLER · Limit Distributions for Sums of Independent Random Vectors: Heavy Tails in Theory and Practice
MENGERSEN, ROBERT, and TITTERINGTON · Mixtures: Estimation and Applications
MICKEY, DUNN, and CLARK · Applied Statistics: Analysis of Variance and Regression, *Third Edition*
* MILLER · Survival Analysis, *Second Edition*
MONTGOMERY, JENNINGS, and KULAHCI · Introduction to Time Series Analysis and Forecasting
MONTGOMERY, PECK, and VINING · Introduction to Linear Regression Analysis, *Fifth Edition*
MORGENTHALER and TUKEY · Configural Polysampling: A Route to Practical Robustness
MUIRHEAD · Aspects of Multivariate Statistical Theory
MULLER and STOYAN · Comparison Methods for Stochastic Models and Risks
MURTHY, XIE, and JIANG · Weibull Models
MYERS, MONTGOMERY, and ANDERSON-COOK · Response Surface Methodology: Process and Product Optimization Using Designed Experiments, *Third Edition*
MYERS, MONTGOMERY, VINING, and ROBINSON · Generalized Linear Models. With Applications in Engineering and the Sciences, *Second Edition*
NATVIG · Multistate Systems Reliability Theory With Applications
† NELSON · Accelerated Testing, Statistical Models, Test Plans, and Data Analyses
† NELSON · Applied Life Data Analysis
NEWMAN · Biostatistical Methods in Epidemiology
NG, TAIN, and TANG · Dirichlet Theory: Theory, Methods and Applications
OKABE, BOOTS, SUGIHARA, and CHIU · Spatial Tesselations: Concepts and Applications of Voronoi Diagrams, *Second Edition*
OLIVER and SMITH · Influence Diagrams, Belief Nets and Decision Analysis
PALTA · Quantitative Methods in Population Health: Extensions of Ordinary Regressions
PANJER · Operational Risk: Modeling and Analytics
PANKRATZ · Forecasting with Dynamic Regression Models
PANKRATZ · Forecasting with Univariate Box-Jenkins Models: Concepts and Cases
PARDOUX · Markov Processes and Applications: Algorithms, Networks, Genome and Finance
PARMIGIANI and INOUE · Decision Theory: Principles and Approaches
* PARZEN · Modern Probability Theory and Its Applications
PEÑA, TIAO, and TSAY · A Course in Time Series Analysis
PESARIN and SALMASO · Permutation Tests for Complex Data: Applications and Software

*Now available in a lower priced paperback edition in the Wiley Classics Library.
†Now available in a lower priced paperback edition in the Wiley–Interscience Paperback Series.

PIANTADOSI · Clinical Trials: A Methodologic Perspective, *Second Edition*
POURAHMADI · Foundations of Time Series Analysis and Prediction Theory
POURAHMADI · High-Dimensional Covariance Estimation
POWELL · Approximate Dynamic Programming: Solving the Curses of Dimensionality, *Second Edition*
POWELL and RYZHOV · Optimal Learning
PRESS · Subjective and Objective Bayesian Statistics, *Second Edition*
PRESS and TANUR · The Subjectivity of Scientists and the Bayesian Approach
PURI, VILAPLANA, and WERTZ · New Perspectives in Theoretical and Applied Statistics
† PUTERMAN · Markov Decision Processes: Discrete Stochastic Dynamic Programming
QIU · Image Processing and Jump Regression Analysis
* RAO · Linear Statistical Inference and Its Applications, *Second Edition*
RAO · Statistical Inference for Fractional Diffusion Processes
RAUSAND and HØYLAND · System Reliability Theory: Models, Statistical Methods, and Applications, *Second Edition*
RAYNER, THAS, and BEST · Smooth Tests of Goodnes of Fit: Using R, *Second Edition*
RENCHER and SCHAALJE · Linear Models in Statistics, *Second Edition*
RENCHER and CHRISTENSEN · Methods of Multivariate Analysis, *Third Edition*
RENCHER · Multivariate Statistical Inference with Applications
RIGDON and BASU · Statistical Methods for the Reliability of Repairable Systems
* RIPLEY · Spatial Statistics
* RIPLEY · Stochastic Simulation
ROHATGI and SALEH · An Introduction to Probability and Statistics, *Second Edition*
ROLSKI, SCHMIDLI, SCHMIDT, and TEUGELS · Stochastic Processes for Insurance and Finance
ROSENBERGER and LACHIN · Randomization in Clinical Trials: Theory and Practice
ROSSI, ALLENBY, and McCULLOCH · Bayesian Statistics and Marketing
† ROUSSEEUW and LEROY · Robust Regression and Outlier Detection
ROYSTON and SAUERBREI · Multivariate Model Building: A Pragmatic Approach to Regression Analysis Based on Fractional Polynomials for Modeling Continuous Variables
* RUBIN · Multiple Imputation for Nonresponse in Surveys
RUBINSTEIN and KROESE · Simulation and the Monte Carlo Method, *Second Edition*
RUBINSTEIN and MELAMED · Modern Simulation and Modeling
RYAN · Modern Engineering Statistics
RYAN · Modern Experimental Design
RYAN · Modern Regression Methods, *Second Edition*
RYAN · Statistical Methods for Quality Improvement, *Third Edition*
SALEH · Theory of Preliminary Test and Stein-Type Estimation with Applications
SALTELLI, CHAN, and SCOTT (editors) · Sensitivity Analysis
SCHERER · Batch Effects and Noise in Microarray Experiments: Sources and Solutions
* SCHEFFE · The Analysis of Variance
SCHIMEK · Smoothing and Regression: Approaches, Computation, and Application
SCHOTT · Matrix Analysis for Statistics, *Second Edition*
SCHOUTENS · Levy Processes in Finance: Pricing Financial Derivatives
SCOTT · Multivariate Density Estimation: Theory, Practice, and Visualization
* SEARLE · Linear Models
† SEARLE · Linear Models for Unbalanced Data
† SEARLE · Matrix Algebra Useful for Statistics
† SEARLE, CASELLA, and McCULLOCH · Variance Components
SEARLE and WILLETT · Matrix Algebra for Applied Economics
SEBER · A Matrix Handbook For Statisticians
† SEBER · Multivariate Observations

*Now available in a lower priced paperback edition in the Wiley Classics Library.
†Now available in a lower priced paperback edition in the Wiley–Interscience Paperback Series.

SEBER and LEE · Linear Regression Analysis, *Second Edition*
† SEBER and WILD · Nonlinear Regression
SENNOTT · Stochastic Dynamic Programming and the Control of Queueing Systems
* SERFLING · Approximation Theorems of Mathematical Statistics
SHAFER and VOVK · Probability and Finance: It's Only a Game!
SHERMAN · Spatial Statistics and Spatio-Temporal Data: Covariance Functions and Directional Properties
SILVAPULLE and SEN · Constrained Statistical Inference: Inequality, Order, and Shape Restrictions
SINGPURWALLA · Reliability and Risk: A Bayesian Perspective
SMALL and McLEISH · Hilbert Space Methods in Probability and Statistical Inference
SRIVASTAVA · Methods of Multivariate Statistics
STAPLETON · Linear Statistical Models, *Second Edition*
STAPLETON · Models for Probability and Statistical Inference: Theory and Applications
STAUDTE and SHEATHER · Robust Estimation and Testing
STOYAN · Counterexamples in Probability, *Second Edition*
STOYAN, KENDALL, and MECKE · Stochastic Geometry and Its Applications, *Second Edition*
STOYAN and STOYAN · Fractals, Random Shapes and Point Fields: Methods of Geometrical Statistics
STREET and BURGESS · The Construction of Optimal Stated Choice Experiments: Theory and Methods
STYAN · The Collected Papers of T. W. Anderson: 1943–1985
SUTTON, ABRAMS, JONES, SHELDON, and SONG · Methods for Meta-Analysis in Medical Research
TAKEZAWA · Introduction to Nonparametric Regression
TAMHANE · Statistical Analysis of Designed Experiments: Theory and Applications
TANAKA · Time Series Analysis: Nonstationary and Noninvertible Distribution Theory
THOMPSON · Empirical Model Building: Data, Models, and Reality, *Second Edition*
THOMPSON · Sampling, *Third Edition*
THOMPSON · Simulation: A Modeler's Approach
THOMPSON and SEBER · Adaptive Sampling
THOMPSON, WILLIAMS, and FINDLAY · Models for Investors in Real World Markets
TIERNEY · LISP-STAT: An Object-Oriented Environment for Statistical Computing and Dynamic Graphics
TSAY · Analysis of Financial Time Series, *Third Edition*
TSAY · An Introduction to Analysis of Financial Data with R
UPTON and FINGLETON · Spatial Data Analysis by Example, Volume II: Categorical and Directional Data
† VAN BELLE · Statistical Rules of Thumb, *Second Edition*
VAN BELLE, FISHER, HEAGERTY, and LUMLEY · Biostatistics: A Methodology for the Health Sciences, *Second Edition*
VESTRUP · The Theory of Measures and Integration
VIDAKOVIC · Statistical Modeling by Wavelets
VIERTL · Statistical Methods for Fuzzy Data
VINOD and REAGLE · Preparing for the Worst: Incorporating Downside Risk in Stock Market Investments
WALLER and GOTWAY · Applied Spatial Statistics for Public Health Data
WEISBERG · Applied Linear Regression, *Third Edition*
WEISBERG · Bias and Causation: Models and Judgment for Valid Comparisons
WELSH · Aspects of Statistical Inference
WESTFALL and YOUNG · Resampling-Based Multiple Testing: Examples and Methods for *p*-Value Adjustment

*Now available in a lower priced paperback edition in the Wiley Classics Library.
†Now available in a lower priced paperback edition in the Wiley–Interscience Paperback Series.

* WHITTAKER · Graphical Models in Applied Multivariate Statistics
WINKER · Optimization Heuristics in Economics: Applications of Threshold Accepting
WOODWORTH · Biostatistics: A Bayesian Introduction
WOOLSON and CLARKE · Statistical Methods for the Analysis of Biomedical Data, *Second Edition*
WU and HAMADA · Experiments: Planning, Analysis, and Parameter Design Optimization, *Second Edition*
WU and ZHANG · Nonparametric Regression Methods for Longitudinal Data Analysis
YIN · Clinical Trial Design: Bayesian and Frequentist Adaptive Methods
YOUNG, VALERO-MORA, and FRIENDLY · Visual Statistics: Seeing Data with Dynamic Interactive Graphics
ZACKS · Stage-Wise Adaptive Designs
* ZELLNER · An Introduction to Bayesian Inference in Econometrics
ZELTERMAN · Discrete Distributions—Applications in the Health Sciences
ZHOU, OBUCHOWSKI, and McCLISH · Statistical Methods in Diagnostic Medicine, *Second Edition*

*Now available in a lower priced paperback edition in the Wiley Classics Library.
†Now available in a lower priced paperback edition in the Wiley–Interscience Paperback Series.